Energy-Based Safety

Despite steady improvements in reducing minor injuries over the past two decades, the rate of serious injuries and fatalities (SIFs) has remained stubbornly flat. Research has revealed that the hazards causing minor injuries are not the same as those leading to fatalities, making traditional safety approaches insufficient. Energy-Based Safety (EBS) presents a science-backed strategy for SIF prevention, built on four core principles that emphasize the role of hazardous energy, the power of effective controls, and the critical influence of workers in shaping safety outcomes.

This book is a useful resource for those committed to saving lives and delivering change in workplace safety through new ways of injury prevention. Divided into two parts, it firstly explores the research behind EBS and investigates each principle with evidence, case studies, and practical insights from real-world implementation. The second part focuses on incorporating EBS into existing safety practices, from design and work planning to incident learning and performance measurement.

Through reading this book, readers will gain practical tools and methods to enhance safety programs, making SIF prevention effective and sustainable in their workplace. Written for safety professionals, *Energy-Based Safety: A Scientific Approach to Preventing Serious Injuries and Fatalities (SIFs)* is an enlightening read for industry leaders in occupational health and safety and practitioners responsible for workplace safety and occupational risk.

Energy-Based Safety

A Scientific Approach to Preventing Serious Injuries and Fatalities (SIFs)

Matthew R. Hallowell

CRC Press
Taylor & Francis Group
Boca Raton London New York

CRC Press is an imprint of the
Taylor & Francis Group, an **informa** business

Designed cover image: Alamini Creative Group

First edition published 2026
by CRC Press
2385 NW Executive Center Drive, Suite 320, Boca Raton FL 33431

and by CRC Press
4 Park Square, Milton Park, Abingdon, Oxon, OX14 4RN

CRC Press is an imprint of Taylor & Francis Group, LLC

ISBN: 978-1-041-07633-9 (hbk)
ISBN: 978-1-041-07590-5 (pbk)
ISBN: 978-1-003-64148-3 (ebk)

DOI: 10.1201/9781003641483

Typeset in Times
by SPi Technologies India Pvt Ltd (Straive)

Dedication

For the men and women who build, power, and sustain our world –your lives are not negotiable.

Contents

Acknowledgments

I am deeply grateful to the exceptional professionals I've worked with through the Construction Safety Research Alliance (CSRA) and the Edison Electric Institute (EEI). Their insights, participation, and commitment to advancing the science of safety have shaped this work in essential ways. They made the research process meaningful, rewarding, and genuinely enjoyable.

I also owe a debt to the front-line workers and field supervisors who welcomed us onto their job sites. Their observations shaped our hypotheses, and their workplaces became the laboratories where those ideas were tested. Without their partnership, this book would not exist.

I am also grateful to the many students I have had the privilege to work with over the years. Each has contributed in some way through questions, research, fieldwork, or writing. Their curiosity, discipline, and persistence have left a meaningful imprint on this work.

This book builds on the foundation laid by the researchers and thinkers who came before me. The ideas of Sidney Dekker, Todd Conklin, and Erik Hollnagel have seeped into my subconscious, while the foundational contributions of William Haddon and James Reason continue to anchor much of our empirical research.

I am especially thankful to Brad MacLean, Cassidy MacLean, Siddharth Bhandari, and Matt Morris, who reviewed early (messy) drafts of this book and provided ideas and encouragement when I needed it most. I am also deeply grateful to my beta reviewers: Bryan Adams, Matt Compher, Mike Court, Ryan Edgerton, Elif Erkal, Peter Lineen, Bobby Lu, Al Payton, Ayleen Perez, Carren Spencer, and Mike Quashne. Your diverse perspectives helped refine the book's clarity, applicability, and impact.

Above all, to my wife, Robyn Hallowell who adds the color to my life: thank you for lending your writing talent, your clarity, your love, and your encouragement. You made this book sharper, and you made me better. And to my two sons, Rowan and Jack – without your boundless energy, this book would've been done years ago. But without your curiosity, your snacks, and the gift of letting me disappear into my thoughts, it never would have been finished at all.

Preface

The safety profession has made remarkable progress over the past several decades. Minor injuries have declined, and safety is now a core value in many organizations. This achievement reflects the commitment, ingenuity, and collaboration across industries.

Despite these gains, Serious Injuries and Fatalities (SIFs) remain stubbornly persistent. The tools that successfully reduced minor injuries have not delivered the same results for SIFs. The numbers have plateaued for more than a decade, signaling that SIFs arise from different conditions and require a distinct approach.

This book, *Energy-Based Safety: A Scientific Approach to Preventing Serious Injuries and Fatalities (SIFs)*, is written with that challenge in mind. It reflects more than two decades of research and development. The framework presented here has been shaped by hundreds of colleagues who tested ideas, challenged assumptions, and kept the work grounded in the realities of front-line operations.

At the heart of this book are four principles that define Energy-Based Safety (EBS), offering a clear and structured lens to address SIFs:

1. Energy causes harm
2. More energy causes more harm, and high-energy causes SIFs
3. Controls save lives
4. People-driven performance

For those well-read in safety, EBS is not a revelation. The idea of "following the energy" has long been part of the safety conversation, and many of the concepts in this book have surfaced in prior work. What's different here is the synthesis. This book extends those foundational ideas by linking long-standing theory with the latest science and unifying them into a cohesive, evidence-based framework for SIF prevention.

The goal of this book is not simply to inform, but to inspire action, guide learning, and foster collaboration within the safety community. While the framework and principles are grounded in rigorous research, this is not the final word on SIF prevention. It is a starting point designed to be durable, yet flexible enough to evolve as our understanding deepens.

Science is never finished, and neither is EBS. Even as this book goes to print, the thinking continues to advance. My hope is that you won't just implement EBS, but that you'll help shape what comes next.

Most of all, I hope this book gives you the confidence to act, knowing that your approach is grounded in evidence, not opinion.

WHO THIS BOOK IS FOR

This book is written for safety professionals who want to strengthen their SIF prevention efforts with the latest research. It is for those looking to build on what already works by focusing more deliberately on the hazards that cause the most serious harm. While safety professionals are the primary audience, operational leaders and executives may also find value in understanding how safety systems can evolve to address SIFs more directly. Although the ideas in this book are grounded in research, the goal was to present the evidence in a way that is easy to understand and apply.

STRUCTURE OF THE BOOK

This book is organized into two main parts of roughly equal length, with six chapters each.

Part I lays the theoretical foundation for EBS. It begins by examining why SIFs remain an urgent and unresolved challenge, even in organizations with mature safety systems. The chapters that follow introduce the core principles of EBS and explain how they provide a structured, evidence-based pathway to prevent SIFs. These principles do more than offer a new perspective; they sharpen the focus on hazards that cause the most serious harm and provide a practical framework for addressing them.

Part II turns theory into practice. It shows how the principles of EBS can be embedded into familiar safety activities, including design and planning, pre-job briefs, Safety Walks, safety measurement, and learning. Each chapter offers practical guidance for strengthening what already works. The book concludes with actionable takeaways, including ten steps for integration and key lessons learned from early implementation.

While the book is designed to be read cover to cover, each chapter is designed to stand on its own. Readers are encouraged to navigate by interest or need, whether reading the full text or referencing specific chapters. The standalone format is intentional, allowing the book to serve not only as a guide during initial reading but also as a long-term reference. Readers can return to specific sections as new challenges arise or when particular topics become relevant in their work.

Author Biography

Dr. Matthew Hallowell is the Founder and Executive Director of the Construction Safety Research Alliance and a President's Teaching Scholar at the University of Colorado Boulder, where he holds a chaired professorship in engineering. He has authored more than 100 peer-reviewed journal articles and mentored over a dozen PhD students who now serve as faculty at leading institutions.

Matt's research focuses on serious injury and fatality (SIF) prevention, with a particular emphasis on translating scientific principles into field-deployable methods for high-risk industries. He serves as the principal technical advisor to the Edison Electric Institute, the INGAA Foundation, and other major industry organizations. His work has been recognized with the National Science Foundation CAREER Award and the inaugural Thomas Farrell Safety Leadership and Innovation Award.

Before entering academia, Matt worked in construction as a laborer, project engineer, and quality inspector. That experience informs a research approach grounded in practicality and relevance for those doing the work.

Part 1

EBS Theory

The central challenge of this book is straightforward: traditional safety methods have reduced minor injuries, but they have not moved the needle on serious injuries and fatalities (SIFs). Part I explores why, building the scientific foundation for energy-based safety (EBS) as an evidence-based approach to SIF prevention.

While traditional safety efforts often give hazards roughly equal attention, EBS begins with a different premise: certain hazards – those involving high levels of energy – demand focused effort and stronger safeguards. This section deepens the ideas introduced in the preface by unpacking the science and logic behind them.

Rather than offering abstract theory, the chapters in Part I present research-backed concepts that support practical decisions: how hazards are prioritized, how risk is assessed, and how safeguards are designed. Along the way, readers will be introduced to foundational tools such as the energy wheel, the Stuff That Kills You (STKY), the 1500-Joule threshold, and the rules that distinguish Direct and Alternative Controls.

Figure P1.1 illustrates the roadmap for this section, tracing the development of the four principles that underpin EBS. Each chapter builds on the last, providing a

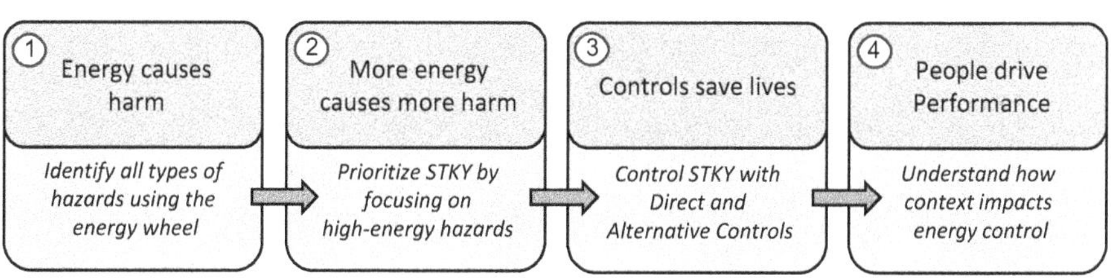

FIGURE P1.1 Four principles of energy-based safety (EBS). Our roadmap for Chapters 3–6.

DOI: 10.1201/9781003641483-1

structured framework for understanding the conditions that give rise to SIFs. By the end of this section, readers will have a durable understanding of these conditions and a clear framework for addressing them with focus and intent.

While readers may be eager to dive into implementation, the chapters in Part I provide the essential foundation for practical application. This groundwork prepares readers for Part II, where these concepts are integrated into core safety activities.

1 Addressing the SIF Plateau

Safety is often measured in injury rates. But behind every number lies a story with lives forever changed, families torn apart, futures rewritten. For those of us in the safety profession, these stories are not abstract; they are the foundation of our work.

Some stories just stick with you. They become symbols of the challenges we face and the urgency of our mission. One such story is that of Juan Rodriguez, a 27-year-old construction worker whose life was forever altered by an avoidable tragedy.

> Juan was known for his diligence and strong work ethic. As the primary earner in his family, Juan supported his wife, Maria, and their two young children, ages four and six. His parents had immigrated to the United States in search of a better life, instilling in Juan a deep sense of responsibility and determination.
>
> On a bright summer morning, Juan was assigned to assist with the groundwork near a new commercial development. The site was bustling with activity. As Juan traversed the site, a miscommunication occurred between the dozer operator and spotter. The operator, under the impression that the area was clear, began reversing the dozer. With his back turned, Juan was struck and run over by the dozer before he could react. The severe crush injuries compromised blood flow and caused tissue damage so extensive that emergency amputation was required to save his life. Juan lost his right leg and experienced chronic pain ever since.
>
> The extent of the injury meant that Juan could no longer work in construction. This has left his family facing significant emotional and financial challenges, as Maria must now care for Juan and their children with limited resources. The small company that employed Juan went out of business 8 months later.

Stories like Juan's serve as a powerful reminder of why our work in safety matters. As the saying goes, whoever saves one life, saves the whole world. The same principle applies to prevention: every serious injury or fatality (SIF) we prevent impacts not just an individual, but ripples through the lives, families, and communities connected to them. To truly understand why these tragedies persist, we need to look beyond stories and into the data.

SIFs are not random occurrences; they result from systems that create hazards and safeguards that neglect to account for human fallibility. These events happen several times a day, even in developed countries. For example, the US construction industry still accounts for nearly 1,000 fatalities a year,[1] a statistic that has remained unchanged for over a decade.

The safety profession has led the way in making remarkable strides in the reduction of injuries. Injury rates across industries have declined dramatically, thanks to

DOI: 10.1201/9781003641483-2

the diligent efforts of safety professionals, leaders, and front-line workers. Traditional safety programs, rooted in the principles of hazard identification and mitigation, have undoubtedly prevented untold suffering. These efforts have shaped a culture where safety is no longer an afterthought but a core value.

Yet, as with any journey, progress reveals new challenges. While we've succeeded in reducing low-severity injuries, our efforts have proven less effective at preventing fatalities. This stark reality highlights a critical challenge: traditional approaches, though effective in many areas, fail to address the unique dynamics of SIFs. Figure 1.1 makes this point unmistakably clear.

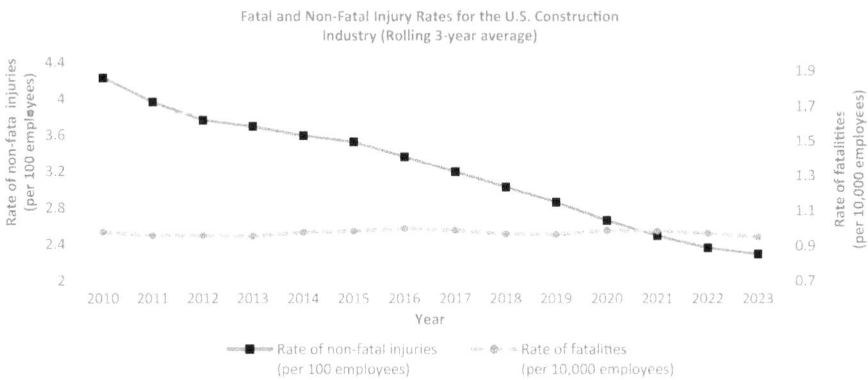

FIGURE 1.1 Fatal and non-fatal injury trends for the US construction industry (2010–2023). While non-fatal injury rates have declined significantly, fatality rates have remained statistically flat over the same period.

Based on fatal and non-fatal injury data from 2010 to 2023 collected by the Occupational Safety and Health Administration (OSHA) and published by the Bureau of Labor Statistics (BLS).

All leaders should see, fully grasp, and appreciate the significance of Figure 1.1. This graph reveals a consistent trend across industries and geographies. While the rate of lower-severity injuries is steadily declining, the rate of fatalities has plateaued. This disconnect appears to stem from an overemphasis on managing recordable injury rates and "generalized" safety programs that treat all hazards with equal emphasis.

I am often invited to discuss SIF prevention at corporate board meetings. While it seems obvious that preventing SIFs is the most important thing we do in safety, I frequently encounter leaders who think otherwise. They argue that all injuries are important, and a SIF-focused strategy might even lead to higher injury rates. When faced with this perspective, I ask a provocative question: "How many recordable injuries would you trade for one life?" Faced with uncomfortable silence, their first instinct, whether they say it out loud or not, is usually "all of them." If this were true, then our safety efforts should focus exclusively on SIF prevention.

However, there are moments when someone pauses, reflects, and provides a different answer. At one meeting, someone offered the number "10,000." This response changed the conversation entirely. It forced everyone to consider what that truly means. If we were to believe that one life is worth 10,000 minor injuries, then for every unit of effort spent managing low-severity injuries, we should be devoting 10,000 times as much effort to preventing SIFs. That's not exaggeration; it is simple logic and math. That means putting vastly more time, attention, and resources into preventing SIFs, not just saying they matter more.

About a decade ago, I had an experience that underscores this point. I was invited to speak at a safety summit for a large multinational company. At that summit, the company proudly celebrated their achievement of reducing their recordable injury rate from 0.60 to 0.45. The atmosphere was jubilant, as if the reduction marked the pinnacle of safety success. Yet, amid the celebration, the company had also experienced four fatalities that year. The stark reality of four fatalities was lost in the celebration of minor injury reduction.

A decade later, I attended the same safety summit, speaking to many of the same people. By this time, their recordable injury rate had declined even further to 0.20. However, the number of fatalities remained the same, with four lives lost once again. The tone of this second summit was markedly different. Gone was the jubilation of the earlier meeting, replaced by solemn acknowledgment that their approach was not addressing the incidents that matter most. It was clear that a new approach was needed.

MISUSE OF THE SAFETY PYRAMID

Just when I think the discussion on SIF prevention is on track, a question often arises, especially among senior board members unfamiliar with modern research: *What about the safety pyramid? Doesn't preventing less severe injuries also prevent SIFs?*

To understand how this misconception took hold, it is important to revisit the origins of the safety pyramid, which has become one of the most recognizable symbols in safety. Introduced by Herbert Heinrich, the pyramid was based on insurance data suggesting a ratio of 300 near hits to 29 minor injuries to 1 major injury (see Figure 1.2). Frank Bird later expanded this theory, proposing that minor and major accidents often share similar causes and that these ratios remain consistent over time.[2] Although neither Heinrich nor Bird claimed the pyramid was causal or that the ratios were universal truths, their work has been widely cited to support the belief that preventing low-severity injuries will automatically reduce high-severity injuries.[3]

The misinterpretation of the safety pyramid has deeply shaped how safety programs are designed and implemented. Because low-severity injuries occur far more frequently than SIFs, they often dominate safety communications, learning initiatives, and discussions. While this may have contributed to a reduction in recordable injury rates, it has not had an equal impact on SIFs. Metrics like total recordable injury rate and the pursuit of "zero-injury" goals have reinforced this misplaced focus. In some cases, it has led to misguided efforts, such as safety stand-downs triggered by minor incidents like a worker twisting their knee while exiting a truck. These actions reduce attention to serious risks and erode credibility with front-line workers.

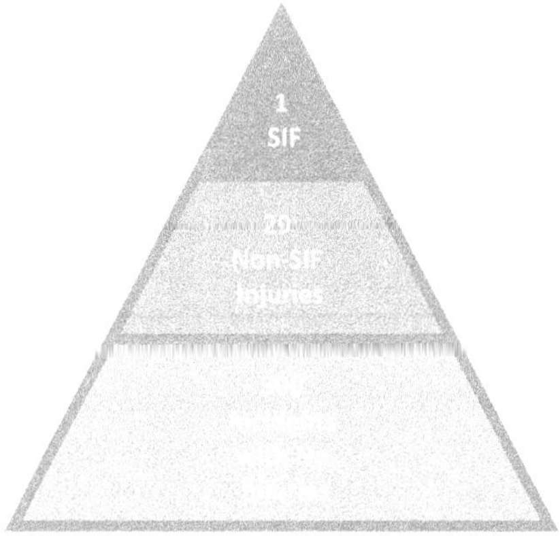

FIGURE 1.2 The traditional safety pyramid and its assumed ratios. This model has been widely discredited due to the assumption that incident ratios are stable across time and context. Empirical evidence shows that the relationship between minor incidents and SIFs is not consistent.

Heinrich, H. W. (1941). *Industrial Accident Prevention: A Scientific Approach*. McGraw-Hill.

The legacy of the safety pyramid exemplifies the perils of making important decisions without evidence. The untested assumption that addressing minor injuries would prevent SIFs has shaped safety programs for generations. It is a cautionary tale about the dangers of accepting untested theory as fact.

To eliminate SIFs, we must move beyond generalized approaches and target the specific hazards that cause the greatest harm. As Marshall Goldsmith said, "What got us here won't get us there." The first step in that transition is establishing a clear definition of what we mean by "SIF."

WHAT IS A SIF (AND WHAT ISN'T)?

Fatality prevention has always been a key priority within the safety community, but the term "SIF" has recently emerged to describe an intentional focus within safety programs. To develop and implement an effective SIF-focused strategy, it is crucial to clearly and consistently define the term SIF. While "fatality" is straightforward, the term "serious" has often been left open to interpretation. As more progressive organizations prioritize SIF prevention and begin to collaborate, achieving clarity and consensus on this definition has become more important than ever.

In 2020, in the Construction Safety Research Alliance (CSRA), we set out to define SIF in a way that would support alignment across companies and industry sectors. Although early definitions provided a starting point, they were vague, which led to inconsistent classifications and confusion.[4] Moreover, these early definitions tended to treat far too many incidents as serious, with lower-impact injuries, such as soft-tissue strains or broken fingers, vastly outnumbering truly high-impact events. This reduces focus and urgency, reinforcing the reality that when everything is labeled as important, nothing is important.

The team's goal was to develop a definition that captured the essence of "serious" in the context of actual injury outcomes. The process was anything but easy, sparking intense debates. For instance, we argued over whether losing the tip of a finger should qualify as serious, a discussion that became far more heated than anyone expected. Without a governed definition to reference, no one was definitively right or wrong; we simply had differing perspectives. In the end, the definition of SIF reflects what the safety community agrees to adopt, govern, and apply consistently. Starting with a blank slate, the challenge was not about finding the correct answer, it was about reaching consensus.

The team held a high bar for defining the word serious. Acknowledging that every injury is serious to the person affected, we had to adopt a more objective, detached approach. To do this, we used the analogy of an emergency room (ER). Everyone in the ER is there for something they consider serious, but the ER nurse must prioritize the most critical cases while others wait. To avoid having too many cases labeled as serious, we needed to take the perspective of the ER nurse, not the patient.

Ultimately, we arrived at a definition of a SIF as a life-ending, life-threatening, or life-altering event. For obvious reasons, we called this definition the LIFE Model. The model defines an injury as a SIF if it results in any of the following outcomes:

1. **Life-Ending:** Loss of life. As the name suggests, this question is simply for the classification of a fatality.
2. **Life-Threatening:** An injury that requires immediate medical intervention or life-saving support to save an injured person's life (e.g., CPR or defibrillation).
3. **Life-Altering:** An injury where the injured worker will not fully recover and will most likely suffer permanent impairment due to the loss of use of a major internal organ, body function, or body part.

Of these criteria, life-altering is the most challenging to apply consistently. The team considered permanent loss as serious because the injured person will never be whole again, a concept that is consistent with the general positions of the UK Health and Safety Executive[5] and the Occupational Safety and Health Administration (OSHA).[6]

Several industry associations have since revisited their definitions of SIFs to improve clarity and consistency. For example, the Edison Electric Institute (EEI) convened a group of safety and medical professionals to identify common examples of life-altering injuries.[7] Their revised criteria, summarized in Appendix 1, reduced the number of events classified as SIFs by 80% by excluding lower-severity injuries like the lost fingertip, temporary soft-tissue injuries, and minor broken bones. This change was widely accepted because grouping minor injuries with catastrophic events weakens focus on the incidents that truly demand attention. By narrowing the definition, we can direct limited resources toward learning from high-impact events.

Creating definitions is not the sexiest task, but it is critically important. The new definition of SIF has improved the consistency of incident classification, which has encouraged companies to collaborate and combine their information into much larger, sector-level datasets. With more data, the research community is beginning to reveal statistically significant insights that were previously hidden.

WHAT IS DIFFERENT ABOUT THE CAUSES OF SIFs?

Once we had defined SIFs, it was time to find out what is different about them. Many people have made claims like, "SIFs happen for different reasons" and "the things that hurt people are not the same as the things that kill people." Although these soundbites are elegant, they are not informative. What are these "reasons" and "things" people keep talking about? Answering this question requires empirical data. Otherwise, we risk replacing one flawed theory with another.

This need for evidence prompted a study aimed at learning, *what, if anything, is different about the causes of SIFs?* If we know these differences, the insights could form the foundation of a truly SIF-focused strategy. While researchers have long studied the causal factors and precursors of fatalities, there had been little investigation into how SIFs differ from lower-severity injuries. As a result, we cannot be certain whether SIF precursors are unique to these high-severity events or simply common factors present in all types of injuries.

Drawing from existing literature in construction, military, psychology, and other fields, we hypothesized which factors might differentiate SIFs from less severe injuries. We gathered injury cases, assessed each for the presence of these factors, and used statistical tests to determine which, if any, were unique to SIFs.

Equipped with a new definition of SIF and dozens of real injury cases, we conducted a controlled experiment.[8] We collected leading data from both low-severity incidents and SIF cases, assessing each for the presence of hypothesized factors. Then we ran statistical tests to determine whether any of those factors were unique to SIFs. The results were surprising. Although most of the hypothesized differentiators were human factors, none were unique to SIFs because they were equally present in SIFs and low-severity cases.

The only differentiator was whether a high-energy hazard was controlled or not. In other words, the "things" that are different about SIFs are the hazards themselves, not the human factors. This means that SIFs cannot be addressed specifically by fixing the workforce. Instead, we must seek to identify, understand, and control the energy involved in our work.

This insight, along with 10 years of research, has led to a focused approach to SIF prevention. This approach is founded on the four principles of Energy-Based Safety (EBS), which offer a clear framework for understanding and controlling the unique forces that drive SIFs. In the chapters ahead, the research behind these principles will be unpacked and translated into practical tools. But before we get there, it's worth pausing to reflect on what this all means.

CONCLUSION

The findings in this chapter underscore the need for a strategic and intentional approach to SIF prevention. However, let me be clear: this is not an argument to focus exclusively on SIFs or to disregard other types of hazards. All hazards deserve attention, *but not equal attention.* Hazards with enough energy to make a SIF likely deserve special focus, more time, and higher-order controls. A truly effective safety system should be especially sensitive to SIF risks, with structured practices for their identification, assessment, and control.

Ultimately, this isn't just about redefining metrics or updating programs. It's about preventing what happened to Juan, and countless others, from ever happening again. These are not abstract concepts; they are grounded in real lives and irreversible consequences.

It is also important to acknowledge that this discussion is centered on acute physical injuries. Themes like health, ergonomics, mental health, psychosocial risks, and psychological safety are critical but fall outside the scope of this book. EBS is not a silver bullet, nor would I claim that it addresses every dimension of safety and health. Instead, it is a framework designed specifically to prevent acute physical injuries, which have been the focus of our research and industry collaboration.

With this caveat in mind, we now turn to the fundamentals of energy in safety. Understanding energy is the foundation of EBS and the key to designing systems that effectively prevent SIFs.

KEY TAKEAWAYS

1. While the rate of low-severity injuries declined significantly in the past 20 years, the rate of fatalities has been statistically stable.
2. The ratios in the safety pyramid are not stable, and a generalized approach to safety is not equally effective at preventing SIFs.
3. SIFs are defined as life-changing, life-altering, or life-ending incidents.
4. The hazards that hurt people are not the same as the hazards that kill people.
5. Controlling high-energy hazards is what separates serious from minor injuries.

NOTES

1 U.S. Bureau of Labor Statistics. (2023). Census of Fatal Occupational Injuries (CFOI) - 2023 data. U.S. Department of Labor. https://www.bls.gov/iif/.

2 Bird, F. E., & Germain, G. L. (1969). Damage Control: A New Horizon in Accident Prevention. American Management Association.

3 See the work of Carsten Busch for a detailed history of Heinrich's work. Busch, C. (2022). Preventing industrial accidents: Reappraising H. W. Heinrich – More than triangles and dominoes. Routledge.

4 Bayona, A., Bhandari, S., Hallowell, M., Sherratt, F., Bailey, J. M., & Upton, J. (2023). What is a serious injury? A model for defining serious injuries & fatalities. *Professional Safety*, 68(09), 22–30.

5 U.K. Health and Safety Executive (HSE). (n.d.). Specified injuries to workers. www.hse.gov.uk/riddor/specified-injuries.htm.

6 Cal/OSHA. (2020). Changes to the definitions of serious injury and illness and reporting to Cal/OSHA (AB 1804 and 1805) www.dir.ca.gov/dosh/Serious-injury-FAQ.html.

7 Edison Electric Institute (EEI). (2020). Serious injury and fatality (SIF) criteria. https://www.eei.org/-/media/Project/EEI/Documents/Issues-and-Policy/Power-to-Prevent-SIF/EEISIF.pdf.

8 Bayona, A., Hallowell, M.R., Bhandari, S. (2024). "The things that hurt people are not the same as the things that kill people: Key differences in the proximal causes of low- and high-severity construction injuries. *Journal of Construction Engineering and Management*, 150(8): 04024089.

2 Introduction to Energy-Based Safety (EBS)

To really understand EBS, we must start with the basics: the physics of energy. Energy is the driving force behind virtually every aspect of our lives. Without it, nothing could move, grow, or function. But just as energy is essential for life and work, it is also the fundamental cause of every SIF.

Put simply, energy is the ability to do work. It drives every task we perform, from moving heavy loads with machinery to cutting materials with hand tools. But when uncontrolled, this same energy causes the most catastrophic workplace events, such as explosions, falling loads, or electrical contact. Energy is both what makes work possible and what makes it dangerous.

To understand the role of energy in safety, we must first understand its two fundamental forms: potential and kinetic. Potential energy is stored energy, such as a tool at height, a cable under tension, or a compressed gas cylinder. Kinetic energy, on the other hand, is the energy of motion. It is visible in moving vehicles, swinging tools, or rotating machinery but also exists in less visible forms like heat (the movement of atoms) or sound (vibrations in the air). Whether visible or hidden, uncontrolled energy presents a hazard that must be identified and managed. Understanding how energy is stored, released, and transferred is the foundation of safety and the first step in preventing SIFs.

ENERGY AS THE FOUNDATION OF SIF PREVENTION

All safety incidents stem from an undesirable transfer of energy to the human body. Viewed through the lens of physics and medicine, safety can be defined as *our ability to harness energy to do work without that energy contacting a human body*. Hazards, in this context, are *relevant sources of energy with the potential to cause harm.*[1] To prevent injuries, we must identify the energy sources in our environment and implement effective controls that prevent transfer to a person.

Research takes the concept a step further. After repeated studies, we now understand that the amount of energy explains the severity of injuries and that our bodies can only withstand a certain amount before permanent harm is sustained.[2] Although other factors matter such as the sharpness of the object and where the energy contacts the body (e.g., the eye is more vulnerable to energy transfer than the thigh), most of injury severity is explained by the amount of energy involved in the incident. Put simply, energy causes harm, and more energy causes more harm.

To my knowledge, in the early 1960s, James J. Gibson was the first to link injuries to the transfer of energy.[3] William Haddon expanded this concept with the energy release theory, which connected ideas from physics, medicine, and safety. He argued

DOI: 10.1201/9781003641483-3

that "damage is caused by the delivery to the body of amounts of energy in excess of the corresponding local or whole-body injury thresholds.[4]"

The implication was clear: injuries can be prevented by controlling the release and transfer of energy. This work was groundbreaking because it reframed injury prevention through the lens of energy control, establishing a scientific and universal framework for understanding safety. In his later work, Haddon outlined the types and layers of safeguards needed to effectively manage energy exposure.[5]

While some attribute these concepts to me, the original ideas belong to Gibson and Haddon. My contribution has been refining their theories, testing hypotheses, and exploring how the concepts work in practice. Without their original creative work, this book would not have been possible.

It may sound bold, but I believe the principles of EBS are so fundamental that they should be considered first principles of safety.[6] These principles provide a universal foundation for SIF prevention that applies to any geography, industry, or task. As with the principles and laws that govern natural sciences, the EBS principles were tested through repeated experimentation, are supported by math and physics, and can be empirically observed and challenged by others.

THE FOUR PRINCIPLES OF ENERGY-BASED SAFETY (EBS)

1. Energy causes harm
2. More energy causes more harm and high energy causes SIFs
3. Controls save lives
4. People drive performance

TRANSLATING EBS INTO EVERYDAY LANGUAGE

While these principles form the foundation of EBS, they don't always land with the people who need them most: the men and women doing the work. To make these concepts more accessible in the field, it is important to translate them into practical, action-oriented terms. The following general overview of EBS theory will serve as a roadmap for the chapters ahead.

Use the Energy Wheel to Look for All Types of Hazards

The principle that energy causes harm resonates with almost any audience, especially when illustrated with relatable examples that are intuitive and easy to grasp. However, when explaining energy, it's often best to avoid overly technical details, such as the nuances of potential and kinetic energy. These discussions tend to appeal only to the most engineering-minded individuals. Instead, I encourage teams to focus on practical, job-relevant categories of energy, including gravity, motion, mechanical, pressure, electrical, chemical, biological, temperature, sound, and radiation. Framing the conversation in this way keeps it grounded in workplace realities and directly applicable to hazard identification.

A powerful tool that makes these concepts actionable is the energy wheel. Often credited to the work of Fleming and Fischer,[7] the energy wheel organizes ten broad energy categories into a visual framework that helps teams identify hazards more comprehensively. This tool, discussed in detail in Chapter 3, has proven invaluable during safety activities such as safety in design and pre-job safety meetings. While some energy sources such as gravity and motion are easily identified by instinct, others such as pressure, mechanical, and chemical energy are often overlooked due to cognitive blind spots that affect even the most experienced workers.[8] The energy wheel counters this tendency by prompting users to consider hazards that might otherwise go unnoticed.

The first step in EBS is recognizing all relevant energy sources in the work environment. The energy wheel helps with this by broadening awareness and ensuring that no hazards are overlooked. Once hazards are identified, the next step is to prioritize those with the greatest potential to cause serious harm. Although the energy wheel itself does not distinguish high-energy hazards, it serves as an essential tool for comprehensive hazard recognition, enabling teams to focus on those with the most significant potential for SIFs.

Prioritize the Stuff That Kills You (STKY)

Research consistently shows that higher levels of energy result in greater harm. At a certain threshold (1500 Joules to be exact), the energy involved becomes so significant that the most likely outcome of contact is a SIF. In the research community, these hazards are referred to as high-energy hazards, but in the field, a simpler term has emerged: the Stuff[9] That Kills You, or STKY. It's catching on because it's simple, sticky, and just edgy enough to get people's attention.

It is essential to provide clear and consistent guidance on what qualifies as STKY and what does not. When left to subjective judgment, STKY classifications often become inconsistent and overly conservative, leading to confusion and labeling too many things as high-risk (i.e., "overclassification"). If everything is deemed STKY, then nothing is truly STKY. The strength of EBS lies in its ability to assess hazards objectively, determining when a hazard is most likely to result in a SIF based on measurable energy levels. Understanding this distinction allows us to move from gut feeling to a method grounded in real evidence. The specific high-energy threshold will be discussed in detail in Chapter 4, along with detailed examples of how to perform energy assessment and practical tools to make these concepts accessible to front-line employees.

Once we've used the energy wheel to identify all hazardous energy and applied the concept of high-energy to find the STKY, we must ensure that the risk of a SIF is minimized.

Make Sure to Have Enough Control

A popular phrase in today's safety language is that safety is the presence of safeguards. Assessing controls helps us put that sentiment into action by objectively answering the question, *when are we safe enough?* This simple question has been challenging to answer. However, with increased understanding of the interplay between engineering and human factors, we have created a set of clear rules that define if the level of protection is adequate or not. For STKY hazards, a control is considered sufficient (and labeled a Direct Control) if it is targeted, mitigates the high-energy hazard, and if it works even if someone makes a mistake.

As my colleague, Mike Quashne, elegantly put it, "A Direct Control will save your life, by itself, even when someone makes a mistake." If we are committed to eliminating SIFs, we must ensure that we have Direct Controls for STKY hazards. In Chapter 5, we will explore the specific criteria for Direct Control and how to evaluate them consistently.

Understand Human Factors and Provide Support

The field of psychology has shown us that human error is not just possible; it is inevitable. This has been translated to the Human and Organizational Performance (HOP) principle that error is normal, and everyone will eventually make a mistake.[10] While we cannot eliminate error, we can strive to understand how work situations influence our ability to identify, assess, and control energy effectively.

As we will explore in Chapter 6, error-prone situations are both predictable and identifiable. For example, work pressure, long hours, and distractions are well-documented factors that increase the likelihood of unintentional slips or lapses in judgment. Although these precursors are not unique to SIFs, repeated studies show they directly impact awareness and decision-making. This is not an issue of "fixing" employees. Instead, it involves understanding the context in which people work and live, recognizing error-prone conditions, and appreciating how human error can undermine our ability to control energy.

Ideally, our safeguards would be so robust and resilient that mistakes could occur without anyone being injured or killed. However, the practical limitations of this ideal must be acknowledged. Almost every physical control is designed, installed, and verified by people. As a result, the reliability of controls is intrinsically tied to human performance. That is, people create and work in systems, and systems drive performance.

EBS AS A WAY TO OPERATIONALIZE MODERN SAFETY PARADIGMS

In this book, I will share the evidence to support the principles of EBS, often at length so you can judge its validity for yourself. I also believe it's important to share the philosophy that shapes my view of safety. That way, you can better understand not just what I'm saying, but where I'm coming from.

Here are a few core beliefs that shape my view of safety. If you're familiar with HOP, Safety II, and the "New View" of safety, many of these will sound familiar.

- Safety is the presence of safeguards
- Error is normal and to be expected; even the best people make mistakes
- Blaming people for mistakes decimates trust and stifles learning and progress

- Error is a symptom of context and never a root cause; context drives behavior
- How we receive bad news will define how much bad news we receive
- Normal, everyday work is rich with learning opportunities

EBS puts these principles into practice by shaping how we measure, manage, and learn from everyday safety activities.

CONCLUSION

At its core, safety is about managing energy. Understanding how energy flows, where it is stored, and how it is controlled is the essence of safety and the foundation for eliminating SIFs. By embracing EBS, we move beyond outdated approaches and focus on the hazards with the greatest potential to cause harm.

Eliminating SIFs is an ambitious but necessary goal. We must strive to create workplaces where no one loses their life or livelihood to uncontrolled energy. The principles and tools of EBS illuminate the path toward this goal, providing a framework to address high-energy hazards and build effective controls. There is a quote from Maya Angelou that captures the spirit of this work: "I did then what I knew how to do. Now that I know better, I do better." EBS represents our opportunity to know better and do better, building a future where safety is defined by the presence of robust controls and is enabled by a shared commitment to eliminating SIFs.

KEY TAKEAWAYS

1. Every injury results from a transfer of energy to the human body; identifying and managing these transfers is at the heart of safety.
2. The energy wheel broadens hazard recognition by helping us identify all forms of energy.
3. More energy means more harm. That's why controlling high-energy hazards (STKY) is the key to preventing SIFs.
4. To be adequate, controls must be targeted and intentional, strong enough to mitigate energy, and forgiving enough to account for human error.
5. Human error is inevitable, and the context in which people work explains how reliably energy will be controlled and managed.

NOTES

1 Albert, A., Hallowell, M. R., Skaggs, M., & Kleiner, B. (2017). Empirical measurement and improvement of hazard recognition skill. *Safety Science*, 93, 1–8.

2 Hallowell, M. R., Alexander, D., & Gambatese, J. A. (2017). Energy-based safety risk assessment: Does magnitude and intensity of energy predict injury severity?. *Construction Management and Economics*, 35(1–2), 64–77.

3 Gibson, J. J. (1961). The contribution of experimental psychology to the formulation of the problem of safety: A brief for basic research. *Behavioral Approaches to Accident Research* (pp. 77–89). New York, NY: Association for the Aid of Crippled Children.

4 Haddon Jr, W. (1963, page 263). A note concerning accident theory and research with special reference to motor vehicle accidents. *Annals of the New York Academy of Sciences*, 107, 635–646.
5 Haddon Jr, W. (1973). Energy damage and the ten countermeasure strategies. *Human Factors*, 15(4), 355–366.
6 A first principle in science is a fundamental, evidence-based truth that serves as the unchanging foundation for understanding and reasoning.
7 Fleming, M. & Fischer, B. (2017). Hazard recognition: Bridging knowledge and competency for process and occupational safety. *Professional Safety*, 62(6), 52–61.
8 Hu, M., Shealy, T., Hallowell, M., & Hardison, D. (2018, April 2–4). Advancing construction hazard recognition through neuroscience: Measuring cognitive response to hazards using functional near infrared spectroscopy [Conference paper]. Construction Research Congress, New Orleans, LA. https://doi.org/10.1061/9780784481288.014.
9 Although sometimes controversial, *Shit* That Kills You is often preferred in the field.
10 Conklin, T. (2019). The 5 principles of human performance: A contemporary update of the building blocks of human performance for the new view of safety. PreAccident Media.

3 Recognizing Hazards using the Energy Wheel

Principle: Energy causes harm

Practice: Identify all types of hazards using the energy wheel

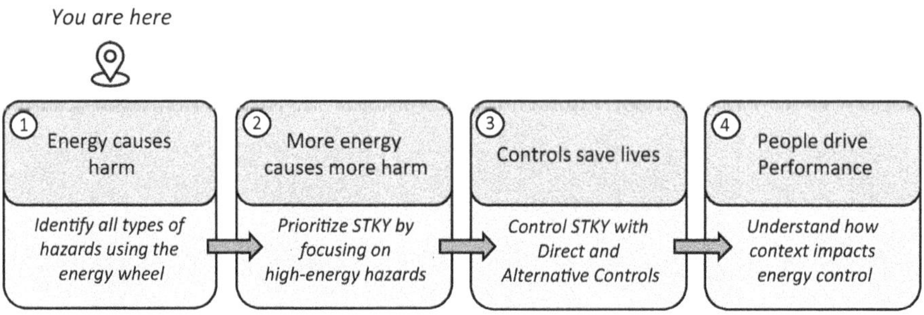

Energy-Based Safety (EBS) begins with identifying all types of energy. While the previous chapter explored the fundamental role of energy in safety, this chapter dives into a practical starting point: using the concept of energy to identify all the relevant hazards around us. Although it may be tempting to dive straight into a focus on SIFs, we must first ensure that we have a complete inventory of hazardous energy.

Left to our own devices, we predictably overlook certain types of energy, even some that are critically important. As we will discuss, the energy wheel addresses this limitation by encouraging us to consider energy sources that might otherwise go unnoticed. The energy wheel is a simple visual tool with 10 hazardous energy types organized into a circle to assist with the comprehensive identification of hazards. By using the energy wheel to identify all potential hazards, we lay the groundwork for a focused approach to SIF prevention.

In this chapter, I will take you on our 15-year journey of studying hazard recognition, exploring how we answered key questions: How good are we at identifying hazards? What hazards do we instinctively recognize, which do we miss, and why? Most importantly, how can we improve? The goal is to leave you with a deeper understanding of how people perceive danger, and why the energy wheel has proven to be such an effective tool despite its simplicity.

Long before I began studying EBS, I worked as a laborer in construction. Because I was young and inexperienced, I was assigned the challenging and dangerous work. The only thing resembling safety on this job was the occasional reminder to be careful. I paid attention to the things that made me nervous, like climbing a ladder to remove old gutters or working under a beam that wasn't fully secured. My hazard recognition was my instinct, and my sense of fear kept me safe.

DOI: 10.1201/9781003641483-4

We have come a long way in safety, but not far enough. We still rely mainly on instinct when recognizing hazards. We identify, discuss, and record whatever hazards come to mind first, while systematically (and unknowingly) missing the rest. If you want evidence of this trend, grab a pile of pre-job briefing forms. You will find that the hazards recorded are mostly the same: slips and trips, pinch points, dropped objects, and other common terms that reflect a repetitive, surface-level discussion. Even among high-performing organizations, we have found that hazard recognition performance is far lower than imagined. Without help, we remain vulnerable to our human limitations.

The energy wheel can be the antidote. Research has uniformly shown that it improves hazard recognition skill, invigorates pre-job planning meetings, and leads to better hazard control. By reminding us to search for hazards that we otherwise overlook, it makes hazard recognition easier and more complete. When front-line workers use the energy wheel for the first time, the most common reaction is, "I wish I had this years ago." I couldn't agree more.

Before we discuss the energy wheel, we must first unpack the role of hazard recognition in safety and examine how it relates to other concepts like risk perception and risk tolerance.

HAZARD RECOGNITION AS A FUNDAMENTAL SKILL

We rely on strong hazard recognition for nearly every tactical safety activity. Front-line workers identify hazards in pre-job safety briefings, supervisors scan for hazards during safety observations, and architects and engineers should anticipate hazards when considering safety in design. If our safety management activities are predicated on this skill, we better be excellent at it, otherwise the weakness will infiltrate the entire safety system.

Most scientific understanding of hazard recognition lies in the branch of applied psychology known as situational awareness. Situational awareness is the process of perceiving an important stimulus, understanding its meaning, and anticipating outcomes.[1] Translated to the context of safety, this process involves: (1) using our senses to recognize the presence of a danger (hazard recognition); (2) judging the level of danger posed by the hazard (i.e., risk perception); (3) feelings about the hazard (risk tolerance); and (4) deciding how to interact with the hazard (i.e., action) (Figure 3.1).[2]

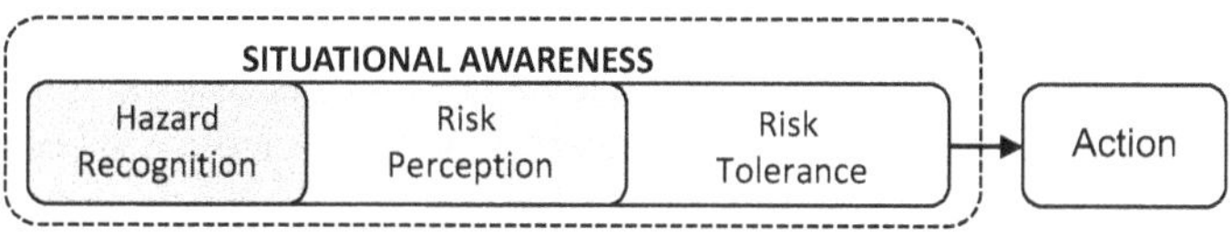

FIGURE 3.1 Situational awareness model highlighting hazard recognition as the first step. Hazard recognition forms the foundation of situational awareness, enabling subsequent steps such as risk assessment and decision-making.

This figure appeared in: Hallowell, M. R. (2021). The energy wheel: The art & science of energy-based hazard recognition. *Professional Safety*, 66(12), 27–33. Reprinted with permission.

Interestingly, the academic community has created a wealth of understanding about risk perception, risk tolerance, and safety behavior, but until recently we knew very little about how we identify a hazard in the first place. This is critical knowledge because risk perception, risk tolerance, and actions are irrelevant if the target hazard isn't identified.

For years, we assumed – without evidence – that we are reasonably good at recognizing hazards. In fact, most traditional safety programs are designed based on three *false* assumptions:

1. If a hazard is present, we can see it.
2. If a hazard will emerge during the work period, we can anticipate it before work begins.
3. Employees overlook hazards because they are distracted, complacent, or poorly trained.

Now that these assumptions have been thoroughly debunked,[3] we need to revisit the very foundations of safety.

THE ENERGY WHEEL

The first principle of EBS is that energy causes harm. Therefore, when it comes to recognizing hazards, our goal should be to see all relevant energy sources in our environment. Since energy exists in various forms, we must be able to scan broadly and consider all energy types, even those that are subtle or complex. For example, we must be able to see obvious sources of energy such as suspended loads or moving vehicles and relatively complex hazards such as cable tension or pressure changes. The energy wheel helps us scan more broadly by aiding our memory and organizing our thinking.

The energy wheel serves as a systematic reminder to scan broadly, helping us overcome the human tendency to focus only on obvious hazards. The energy wheel includes ten types of energy commonly found on worksites (Figure 3.2). Although there are technically only two forms of energy (kinetic and potential), the energy wheel is about practicality, not precision. Using common terms such as gravity, motion, electrical, and pressure makes it easier to think about different types of energy and how they might appear during work.

The energy wheel is simple to use. First, scan your work environment for hazards you see by instinct. Once you have identified all hazards that come to mind, grab the energy wheel, and fill in the gaps. Ask yourself (and your team): What other gravity hazards might we encounter? How about motion, mechanical, electrical, pressure, and so on? These prompts help us think more broadly by tapping into a different part of the brain.[4]

The energy wheel has become one of the most recognized safety innovations. It can be seen on signs and posters, safety vests, and hardhats. It is also plastered on social media and appears to be one of the few safety tools that isn't rife with controversy and debate. Although the energy wheel was originally tested in the construction industry, we have found the ten energy categories to be universal. Whether you are on a construction site, in a hospital, or on the moon, energy transfer is still what causes injury, and the same ten categories are relevant. Definitions and examples of these energy sources are provided in Table 3.1.

Although we are now aware that the energy wheel causes dramatic improvements in hazard recognition, we did not immediately recognize its potential. After all, it is merely ten energy icons organized into a circle. How can something so simple be so transformative? It was through a series of fortuitous experiments that we began to understand its immense impact. To truly understand the energy wheel, you must understand the research behind it. Otherwise, it is easy to dismiss as just another fleeting safety campaign or a form of safety propaganda.

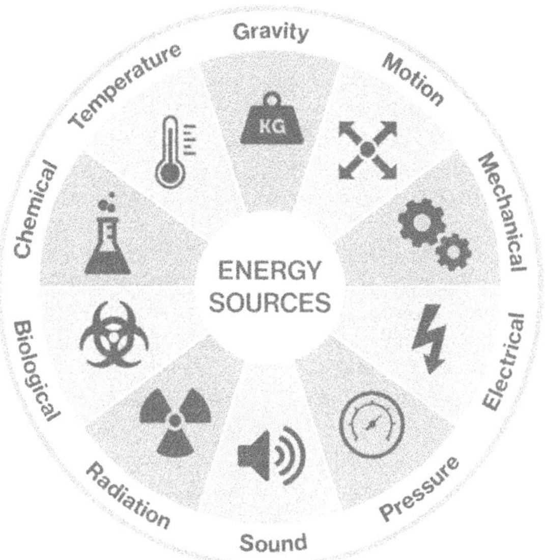

FIGURE 3.2 The energy wheel. A visual tool with ten hazardous energy types organized into a circle to assist with the comprehensive identification of hazards.

This figure appeared in: Hallowell, M. R. (2021). The energy wheel: The art & science of energy-based hazard recognition. *Professional Safety*, 66(12), 27–33. Reprinted with permission.

ENERGY WHEEL RESEARCH: PUSHING THE BOUNDARIES OF SAFETY SCIENCE

My journey with the energy wheel began serendipitously. In my first few years as a professor, I was fortunate to receive funding from the Construction Industry Institute (CII) to study hazard recognition in partnership with an excellent researcher and soon-to-be friend, Dr. Brian Kleiner, from Virginia Tech.

The project wasn't our idea. It was nominated and supported by a cohort of industry professionals who voted for the topic over a dozen others. Along with funding, the

TABLE 3.1
Definitions and Examples of the Ten Energy Categories in the Energy Wheel

Energy Category	Definition	Examples
Gravity	Force caused by the attraction of mass to the earth.	Uneven work surface, work at height, unsecure materials, overhead support structures
Motion	Change in the physical position or location of objects or substances.	Traffic, mobile equipment, projectiles, dust particles
Mechanical	Working parts of a machine or assembly, including rotation, vibration, tension, or compression	Auger, cable, chain fall, angle grinder, gears, pulleys
Electrical	Presence of electrical charge or current	Wires, power lines, power tools, extension cords, transformer, relay
Sound	Audible vibration caused by the contact of two or more objects	Heavy machinery, pile driving, power tools, nail gun
Pressure	Liquid or gas compressed or under vacuum	Pneumatic tire, piping system, tank, hydraulic lines
Temperature	Intensity of heat in an object or substance	Friction, engines, sudden pressure change, steam
Chemical	Toxic objects or substances that pose health risks	Solvents, engine exhaust, silica, wood dust, liquid concrete
Radiation	Objects or substances that emit electromagnetic waves or subatomic particles	Welding, sun exposure, x-ray testing, radioactive waste
Biological	Living organisms or viruses	Bees, snakes, alligators, bears, restrooms

This table appeared in: Hallowell, M. R. (2021). The energy wheel: The art & science of energy-based hazard recognition. *Professional Safety*, 66(12), 27–33. Reprinted with permission.

project came with 20 highly experienced safety professionals who volunteered to participate. This was my dream project as a fledgling academic.

Our project started by understanding the motivations of the team and having conversations about hazard recognition. In one of our earliest meetings, a safety leader from a major construction firm said, "You know, we do all this training, all these observations … and still, people are missing hazards right in front of them. It doesn't make sense."

Heads around the table nodded. These were companies known for their mature safety systems, industry leaders by any standard. But everyone had the same story: Well-trained, engaged employees were overlooking hazards that later contributed to serious incidents.

The word "complacency" came up again and again as the default explanation in incident investigations. If a hazard was obvious in hindsight, the assumption was that the worker must have seen it and chosen to ignore it. After all, a hazard is painfully obvious in retrospect.

But as the group shifted its focus from investigation to learning, the tone began to change. Instead of asking "Why did they make that choice?", they started asking

"What were they actually seeing in the moment?" and "What was happening around them?" This shift made space for a deeper insight.

We began noticing a pattern: the hazards being missed weren't just by new hires or disengaged workers, they were being missed by people with years of experience who cared deeply about their work. Maybe the issue wasn't complacency, but the limits of human perception.

That realization shaped our work. We weren't just studying hazard recognition; we were challenging the assumptions behind decades of investigations. If strong teams were missing hazards for reasons beyond their control, then it wasn't just about accountability and trying to convince them to try harder or care more. It was about understanding which hazards people see, which they miss, and why.

FIELD EXPERIMENTS

To better understand hazard recognition, we employed a rigorous experimental method called multiple baseline testing (MBT). Introduced to me by Dr. Helen Lingard, MBT originated in pharmaceutical research for drug testing, and has since proven effective for testing safety interventions in project-based work. In our study, we used it to understand baseline hazard recognition skills and measure the impact of several potential hazard recognition interventions.[5]

MBT experiments are the gold standard because they allow you to isolate and assign an impact directly to the thing you are testing. That is, you can conclude that a strategy *caused* an improvement in hazard recognition, above and beyond anything else. This is simply not possible with opinions, anecdotes, correlational data, or "micro-experiments."[6]

Our MBT study had three major phases:

1. Pre-Implementation Baseline: Work crews were shadowed to observe and measure hazard recognition skill before any intervention was introduced.
2. Intervention: At randomly selected times during the study period, a new tool was introduced to the work crews that we hypothesized would improve hazard recognition skill.
3. Post-Implementation Baseline: Using the same methods as the baseline phase, the crews were observed after the intervention was introduced to measure changes in hazard recognition skill.

At first, it didn't dawn on us to study the energy wheel as an intervention. Instead, we planned to test three strategies: improving engagement during pre-job safety meetings, visual cues on a large summary chart, and training in virtual reality. Because the experiments were focused on improving hazard recognition, we needed a precise definition of a hazard and a defendable way of measuring skill. In an early meeting, our colleagues from Chevron introduced our team to a new way of thinking about hazards known as the energy wheel. I had not seen it before but found it appealing because it redefined a hazard from just "a danger" to something more specific: a source of energy. It also provided tangible examples that related to field operations. As we designed the study, the energy wheel became an integral component of each intervention and a way to add precision to our measurement of hazard recognition skill.

Results: A Collection of Surprising Insights

The field experiments revealed an interesting and unexpected result: all three interventions improved hazard recognition by the same amount (about 30%).[7] As we later learned, the energy wheel, not the intended interventions, caused the improvements. The interventions were the vessel, but the energy wheel was the actual change agent.

Because there was very little prior research into hazard recognition, the MBT study allowed us to do more than test interventions. For the first time, we empirically measured hazard recognition skill, discerned patterns, and uncovered the fundamental drivers of improvement. Our key findings are summarized in what follows.

Finding 1: We Only Identify about 45% of the Hazards during Our Work Planning

Before an intervention was tested with a crew, we measured their baseline hazard recognition skills. To do this, our graduate student, Alex Albert (now Dr. Alex Albert, tenured professor), visited active worksites to collect data. He was assigned to a crew for a 2-to-3-week period where he observed the daily pre-job briefs and recorded every hazard that the crew discussed or documented. During each work period, he shadowed the crew and recorded all the hazards that they actually faced. This inventory allowed us to measure baseline hazard recognition skill using the simple formula below.

Measuring hazard recognition (HR) skill

$$\text{HR skill}(\%) = \frac{\text{Number of Hazards Identified Prior to Work}}{\text{Number of Hazards Encountered During Work}} \tag{3.1}$$

After observing dozens of work periods over thousands of worker-hours, Alex returned to our team to share the results. **The crews from these industry-leading companies were identifying fewer than half of the hazards**. The reported statistic (45% skill) was shocking enough that we didn't believe it at first. After more interrogation and careful follow-up work, we confirmed that the skill was, indeed, 45% on average. This number was later validated by numerous studies conducted by other researchers in different geographies and trades.

Finding 2: Hazard Recognition Skills Are the Same Regardless of Age, Experience, or Other Personal Factors

We assumed seasoned workers would identify more hazards than their junior counterparts, but we were wrong. There was no statistical relationship between experience and performance. Although experienced workers were better at explaining or responding to hazards, they were just as likely to miss them in the first place.

To test this further, we evaluated the hazard recognition skills of our own team. We went to an active site and followed the same blind field protocol Alex had used with the crews. This wasn't a casual walk; we were focused, deliberate, and motivated to prove ourselves. After all, who's going to try harder to spot hazards than a group of safety professionals being tested on the very thing they teach? Still, we scored just under 50%.

It was humbling and confusing. The people who teach hazard recognition missed hazards at the same rate as those they train. The takeaway was clear: if even top performers consistently miss half the hazards, the problem isn't that people don't care. It must be something deeper.

Finding 3: Not Every Hazard Is Reasonably Identifiable in a Pre-job Safety Briefing

When shadowing crews, it became apparent that hazards are missed for one of two reasons: (1) the hazard was detectable before work began, but the team did not identify it; or (2) the hazard wasn't reasonably detectable with the information available at the time. For example, crews often miss hazards such as pressure vessels and cable tension that are integral components of the planned work. However, other hazards are missed because they emerge from unforeseen change (e.g., a subcontractor unexpectedly staging materials above a workspace as work progresses). It is simply not possible to anticipate every hazard at the very beginning of the work period.

This is described as what is commonly called the "black-line, blue line" problem. The idea, as shown in Figure 3.3, is that work as planned can be different from work as performed. Although this is commonly discussed as the gap between management's vision of work and the real world, we also found that work tends to drift for front-line employees. Because work rarely goes exactly as planned, about 30% of hazards are not reasonably identifiable until they are encountered during work.

FIGURE 3.3 Work as imagined versus work as performed. Roughly 30% of hazards are not identifiable before work begins, underscoring the need for ongoing hazard recognition during work.

This figure appeared in: Hallowell, M. R. (2021). The energy wheel: The art & science of energy-based hazard recognition. *Professional Safety*, 66(12), 27–33. Reprinted with permission.

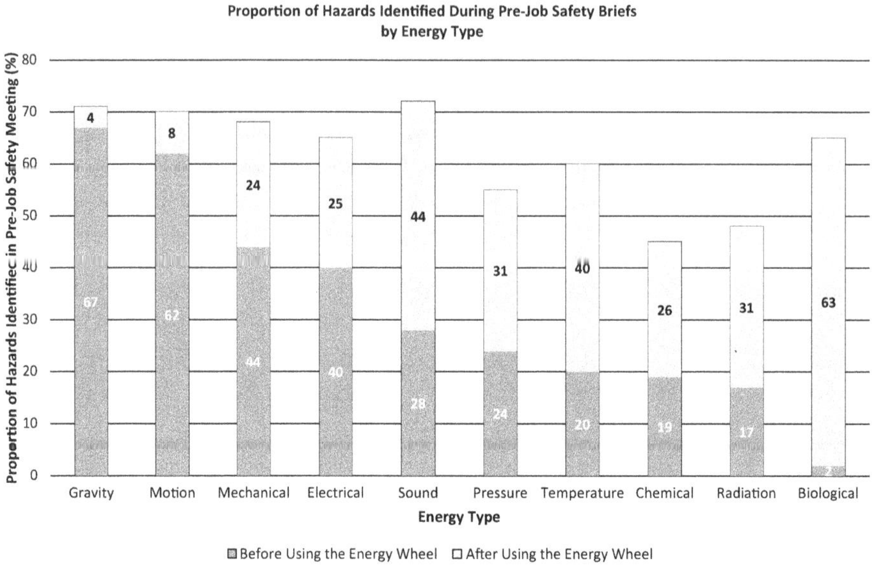

FIGURE 3.4 Impact of the energy wheel on hazard recognition by energy type. Dark gray bars show hazards identified before the energy wheel was introduced; light gray bars show additional hazards identified with it. Total bar height reflects the improved recognition achieved using the energy wheel.

Finding 4: We Tend to See Gravity and Motion Hazards but Overlook Others such as Pressure, Temperature, Radiation, Chemical, and Biological

The patterns in hazard recognition were puzzling. Understanding that the ceiling for pre-job hazard recognition is about 70%, crews were nearly perfect at identifying gravity and motion hazards. They didn't need extra reminders, tools, or training, they just saw them. However, pressure, temperature, radiation, chemical, and biological hazards were overlooked at a comparatively high rate (see Figure 3.4). Why was there such a consistent disparity? This perplexing phenomenon was later explained by our laboratory experiments.

Finding 5: The Energy Wheel Improves Hazard Recognition by an Average of 30%

As interventions were introduced during the experiment, crews were provided with some very basic instruction about the energy wheel through a 60-minute training. They were told that a hazard is *a source of energy that could cause injury, illness, or death* and were instructed to use the energy wheel so that their scan was complete. They were then given the energy wheel shown in Figure 3.2 as a take-away.

As it became apparent that the energy wheel was the real prize, we tested it in isolation without the other surrounding tools (e.g., without the accompanying virtual training experience). As hypothesized, the energy wheel alone improved hazard recognition by over 30%. The simple set of reminders helped crews think differently, and this translated to immediate improvements in skill.

All the evidence pointed to one thing: the energy wheel was working very well. Since we didn't know exactly why, we went to the lab to find out.

HOW BIG OF AN IMPROVEMENT IS 30%?

Our research shows that the average crew identifies and discusses seven hazards per pre-job meeting, many of which are repeated from memory (slips and trips, pinch points). Since the energy wheel improves hazard recognition by 30%, the average crew will then identify and discuss an average of nine hazards per day (i.e., they plan for two additional hazards that they would have otherwise overlooked).

Take a company with 500 employees, and an average crew size of 6 employees. This company would have about 83 crews doing a pre-job once a day for roughly 250 working days per year. With the energy wheel, this company would identify and plan for about 41,500 *additional* hazards per year (250 days × 83 crews × 2 additional hazards per day).

LABORATORY EXPERIMENTS AND RESULTS

The field experiments left us with a sense of curiosity. Since trends in hazard recognition transcended experience and background, we suspected that biological drivers were at play. We were eager to learn why some hazards were identified so easily and others were not, and why the energy wheel was so effective.

Partnering with colleagues across disciplines, we conducted a set of laboratory experiments to complement our field work. We hypothesized that hazards were processed in one part of the brain (the frontal lobe) and that hazards that were more complex would be missed because they require more mental effort to process. We were partially correct.

Our hypothesis was founded on the knowledge that our brains are hardwired to be efficient.[8] Sending and processing electrical signals across the brain's circuits requires massive amounts of glucose and oxygen. Despite its small size, the brain consumes about 20% of our calories each day.[9] When presented with multiple stimuli, the brain conserves energy by focusing on information that is the least expensive to process. Therefore, when presented with a hazard recognition task, neuroscience suggests that the brain will attend to the hazards that require the fewest calories to process. Put simply, we thought our brains subconsciously ignore high-cost hazards to be efficient.

To test our hypothesis, we ran a series of simple experiments that leveraged complex technology. We used noninvasive optical imaging techniques such as functional near infrared spectroscopy (fNIRS) to measure changes in hemoglobin and oxygen concentrations within the brain during a hazard recognition task.[10] With this data, we could tell which part of the brain was working and how hard.

Yet again, the results were surprising. The scans indicated that not one, but two parts of the brain were responsible for hazard recognition: the amygdala and the cerebrum. Commonly identified hazards such as gravity and motion were processed immediately and with very low resource demands. This suggests that gravity and motion hazards are processed in the amygdala, the location of the brain responsible for fight or flight response.[11] Activation of the amygdala is also associated with emotional and physiological responses because the amygdala sends a distress signal to the hypothalamus, which triggers the nervous system to release adrenaline. Psychologists often refer to this as the fast brain (or System I) because it is quick, automatic, and efficient.[12]

Alternatively, hazards that are commonly missed (e.g., mechanical, pressure, and chemical) require much longer processing times and elevated oxygen demand, suggesting that they are processed by the cerebrum, a high-cost part of the brain that is generally responsible for memory, sequencing, and complex problem solving. This is known as the slow brain (or System II) because it is comparatively slow and expensive.

Perhaps obvious in retrospect, our field experiments, laboratory studies, and the work of Kahneman and Tversky converged on a shared insight: the amygdala is responsible for fast, instinctual hazard recognition, while the cerebrum governs slower, more deliberate processing. The distinction between energy types helps explain why some hazards are immediately recognized and others require conscious effort—they engage different parts of the brain. The secret to improving hazard recognition, then, is to deliberately prompt ourselves to consider high-cost hazards we might otherwise overlook. The energy wheel engages this "slow brain" by providing those essential reminders.[13]

CASE EXAMPLE OF WHEN INSTINCT OVERRIDES REASON

During a site visit to a power generation facility in a remote Colorado mountain town, the research came alive for me. As we reached the top floor via elevator, the guide, keeping with tradition, didn't warn me about what was coming. I stepped out onto a metal grating, 250 feet off the ground. Looking down, I could see straight through the perfectly aligned holes, creating the illusion I was standing on air. Although I've never had an issue with heights, my body reacted immediately: vertigo, a sinking feeling in my stomach, flushed face, spiked heart rate, and a few choice words. I knew I was safe because the flooring was engineered, and guardrails surrounded me, but my emotions overrode reason.

What struck me later was how I reacted so strongly to this gravity exposure, but walked right past a tank under extreme pressure without hesitation moments later. Why did one hazard trigger such a visceral response while the other didn't even register? That's when it clicked. My brain had instinctively processed the fall hazard (gravity) through the amygdala, the primitive part of the brain tied to emotion and the fight-or-flight response. My reaction to gravity wasn't logical, it was primal.

We instinctively respond to hazards like gravity and motion because they've been threats to human survival for hundreds of thousands of years. In contrast, industrial hazards like pressure, electrical, and chemical risks are relatively new, emerging with the Industrial Revolution. Our instincts haven't caught up yet. That's where the energy wheel comes in: it reminds us of the dangers our instincts fail to notice.

LESSONS LEARNED AND BEST PRACTICES FOR USING THE ENERGY WHEEL

Over the years, I have collaborated with companies across industries to integrate the energy wheel into their planning processes. By engaging directly with front-line workers, safety leaders, and management teams, we uncovered the following lessons learned and best practices.

1. ***The energy wheel helps us to do what we already do, better.*** Let's face it, the last thing we need is more time-consuming safety practices that take time away from getting the job done. Fortunately, the energy wheel is not an additional safety practice. It is a tool that can be used to support existing practices where identifying hazards is important. This can range from pre-job safety briefs to safety in design reviews.
2. ***Once you learn how to use it, the energy wheel does not take more time.*** Initially, meetings may take slightly longer as teams learn to use the energy wheel, increasing from an average of 10 minutes to about 15 minutes. However, within two weeks of consistent use, discussions become more efficient, returning to the original 10-minute duration. Awkward pauses are replaced by focused reviews of energy categories, and broad questions like "What other hazards are present today?" evolve into targeted ones such as "What chemical hazards might we encounter?" Over time, teams identify and address a greater number and variety of hazards without requiring more time.
3. ***Use instinct first, then use the energy wheel.*** The first step in using the energy wheel is to rely on instinct. Have the team discuss all the hazards that come to mind naturally and then bring out the energy wheel to close the gaps. This approach respects and builds on our ability to recognize roughly half of all hazards with instinct alone. By leveraging our natural strengths and following up with the energy wheel, we create a process that is efficient, thorough, and respectful of workers' innate abilities.
4. ***Sustained integration is the key to lasting impact.*** The effectiveness of the energy wheel depends on how well it is woven into the fabric of the organization's safety system. Visual prompts and quick wins can create early momentum, but they are not enough on their own. Superficial efforts, like a one-time presentation or a passive distribution of stickers may spark initial interest, but they rarely lead to meaningful change. Similarly, relying on a single enthusiastic champion can drive localized success, but the program often fades when that person moves on. Sustainable integration requires

leadership buy-in, systematic training, alignment with safety messaging and activities, and consistent reinforcement. When fully adopted, the energy wheel becomes part of how teams think, talk, and act, ensuring its impact endures across shifts, teams, and turnover.

CONCLUSION

The energy wheel is a simple but powerful tool grounded in the principle that energy causes harm. To prevent that harm, we must first identify all sources of energy in the workplace. While the energy wheel is not specific to SIFs, it ensures that we begin with a complete and comprehensive inventory of hazards. Without this step, we risk overlooking serious hazards that could lead to catastrophic outcomes.

By prompting workers to think systematically about all types of energy, the energy wheel improves hazard recognition by approximately 30%. Its simplicity makes it accessible, while its effectiveness lies in its ability to address cognitive blind spots and expand awareness. The energy wheel bridges the gap between instinctive reactions and deliberate, systematic thinking. It helps workers find more hazards, more easily, without requiring more time.

As foundational as the energy wheel is, it is just the beginning of SIF prevention. While it helps identify all hazards, not all energy poses the same level of threat. The next step is to prioritize the hazards that matter most: high-energy hazards, or the "Stuff That Kills You" (STKY). In the chapters ahead, we will delve deeply into SIF prevention, shifting from broad energy awareness to a targeted focus on controlling these critical hazards.

KEY TAKEAWAYS

1. Although hazard recognition is required for nearly every safety activity, most crews identify fewer than half of the hazards they face during a work period.
2. Hazard recognition skill tends to be the same regardless of age, experience, education, or other personal differences.
3. We instinctually recognize gravity, motion, and mechanical hazards, but we miss other energy types because they require complex thought, reasoning, and problem solving.
4. The energy wheel is a simple tool that provides reminders to scan for categories of hazards that we would otherwise overlook, improving hazard recognition skill by 30%.
5. The energy wheel is based on the understanding that injuries are caused by energy transfer, so hazards are defined as sources of energy that could cause harm.

NOTES

1 Endsley, M. R. (1995). Toward a theory of situation awareness in dynamic systems. *Human Factors*, 37(1), 32–64. https://doi.org/10.1518/001872095779049543.

2 Tixier, A. J. P., Hallowell, M. R., Albert, A., van Boven, L., & Kleiner, B. M. (2014). Psychological antecedents of risk-taking behavior in construction. *Journal of Construction Engineering and Management*, 140(11), 04014052.
3 Albert, A., Hallowell, M. R., Skaggs, M., & Kleiner, B. (2017). Empirical measurement and improvement of hazard recognition skill. *Safety Science*, 93, 1–8.
4 Kahneman, D. (2011). *Thinking, Fast and Slow*. Macmillan.
5 Albert, A., Hallowell, M.R., Kleiner, B., Chen, A., & Golparvar-Fard, M. (2014). Enhancing construction hazard recognition with high-fidelity augmented virtuality. *Journal of Construction Engineering and Management*, 140(7), 04014024. https://doi.org/10.1061/(ASCE)CO.1943-7862.0000860.
6 MBT is a level of rigor that is conspicuously absent in safety. Experimental methods like MBT are rarely used because they require months of controlled field tests that are infeasible for most researchers. They are challenging and expensive, requiring substantial funding and industry collaboration that most researchers cannot access. Fortunately for us, we had time, money, and a large industry team willing to participate.
7 Hallowell, M. R. (2021). The energy wheel: The art & science of energy-based hazard recognition. *Professional Safety*, 66(12), 27–33.
8 Kahneman, D. (2011). *Thinking, Fast and Slow*. Macmillan.
9 Raichle, M. E., & Gusnard, D. A. (2002). Appraising the brain's energy budget. *Proceedings of the National Academy of Sciences*, 99(16), 10237–10239.
10 Hu, M., Shealy, T., Hallowell, M., & Hardison, D. (2018, April 2–4). Advancing construction hazard recognition through neuroscience: Measuring cognitive response to hazards using functional near infrared spectroscopy [Conference paper]. Construction Research Congress, New Orleans, LA. https://doi.org/10.1061/9780784481288.014.
11 Adolphs, R., Tranel, D., Damasio, H., & Damasio, A. R. (1995). Fear and the human amygdala. *Journal of Neuroscience*, 15(9), 5879–5891. https://doi.org/10.1523/jneurosci.15-09-05879.1995.
12 These results are consistent with modern psychology and neuroscience. Nobel prize-winning researcher Daniel Kahneman famously wrote about this in his book, *Thinking Fast and Slow*, where he explained cognition as system I (fast brain) and system II (slow brain). He noted that system I operates automatically and quickly, with little or no effort, and no sense of voluntary control. Alternatively, system II allocates attention to the high-cost mental activities that demand it and requires our conscious direction.
13 In psychology, these types of mental reminders are called mnemonics. You may remember some from your childhood such as PEMDAS for the order of operations in algebra or ROYGBIV for the order of the colors in the rainbow.

4 High-Energy Hazards

How to Find the Stuff That Kills You (STKY)

Principle: More energy causes more harm
Practice: Prioritize the Stuff that Kills You (STKY) by focusing on high-energy hazards

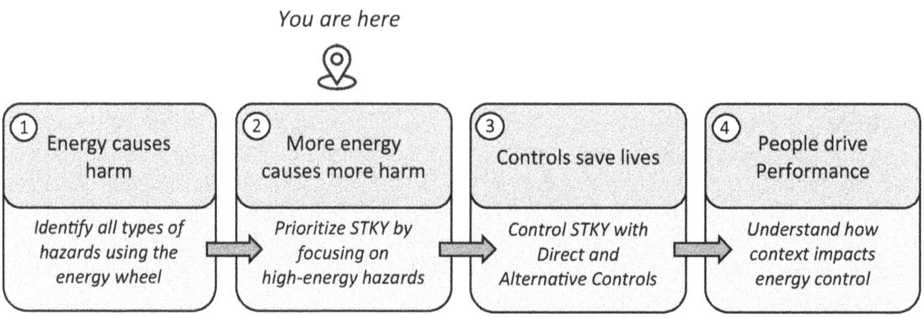

Nothing is more important in the safety profession than saving lives. That is our ultimate purpose. In recent years, the safety profession has sharpened its focus on SIFs, with companies embedding SIF prevention into their corporate messaging, industry groups launching initiatives such as the Edison Electric Institute's *Power to Prevent SIF*, and research organizations like the Construction Safety Research Alliance (CSRA) forming with the express vision of SIF elimination.

But this shift raises a critical question: What must change for SIF prevention to succeed? We already know that traditional safety programs, designed to reduce minor injuries, do not translate to SIF prevention. Research shows that the key to breaking the SIF plateau is recognizing and controlling high-energy hazards. Yet, many safety systems treat all hazards alike, failing to distinguish between those that cause minor injury and those that can kill.

This chapter focuses on that distinction. If we are serious about SIF prevention, we must shift our focus from generally preventing injuries to deliberately controlling high-energy hazards.

 DOI: 10.1201/9781003641483-5

WHY WE NEED AN ANALYTICAL APPROACH TO SIF PREVENTION

On the surface, SIF prevention may seem like a clear path. Can't we simply use our judgment to identify and address serious hazards? Unfortunately, it's not that simple. As psychologists have demonstrated for a century, human judgment of probability and severity is terribly unreliable. When it comes to SIFs, this means we cannot rely on intuition alone to assess risk. We need a structured, data-driven approach. Because high-energy hazards are the driving force behind SIFs, energy assessment is the key.

An objective assessment of hazards begins by asking better questions. For example, asking what was the *worst thing* that could happen always leads to the same conclusion: a SIF. If SIF is not the answer, it means we are not being creative enough. Every situation, no matter how minor, carries the remote possibility of causing a SIF. Infections from small cuts have led to amputations, and falls to the same level have caused severe head trauma. However, not every cut, slip, or trip is "serious."

Instead, we should be asking: *What is the most likely outcome?* This is a simple change in wording, but a transformative change in perspective. It requires a shift from deterministic to probabilistic thinking; from considering one possible outcome to many. We have a difficult time making these assessments with judgment alone, but objective methods can help.

THE PROBLEM WITH USING JUDGMENT TO ASSESS SEVERITY

Before proposing a wholly new approach to hazard assessment, it is essential to fully understand the problem. Human judgment of probability is well known to be significantly influenced by biases, such as overreliance on recent experiences, leading to flawed assessments. The same is true for assessing severity. Although assessments in the extremes are easy (e.g., a massive, suspended load is a serious hazard, and a trip hazard on ground level is not), most hazards are not extreme and obvious; they are slight and tricky.

A simple activity can reveal the challenges of subjectively judging the *most likely* severity of a hazard. Take the image in Figure 4.1, which depicts normal, everyday work. What do you think the most likely outcome would be if the worker were to be injured by the hazard? How confident are you in that assessment?

If you are like me, you relied heavily on personal experience, stories, and gut feelings about the hazard. You thought about what the incident would look like and the different types of injuries that could occur. You may even have thought about the worst possible outcome although specifically encouraged not to. It is easy to fall into these traps, which leaves us guessing.

NOISE EXPERIMENT

To explore severity assessment beyond a few case examples, we conducted a full noise experiment. Noise experiments are simple: you provide the same situation (e.g., Figure 4.1) and measure the extent to which different people provide different assessments.[1] If there is a lot of variability, the process is noisy, and if the assessments

FIGURE 4.1 Worker on a ladder.

are consistent, the process is reliable. Although the assessment of most likely severity seems easy, we correctly hypothesized that there was a great deal of noise in people's assessments.

To conduct the noise experiment, we asked 93 front-line leaders and safety managers to evaluate 25 hazard scenarios. For each scenario, we asked them two questions:

1. What is the most likely injury severity? Selecting among the following four options: first aid, medical case, lost time, or fatality.
2. How confident are you in your response? Using a 1 to 5 scale, ranging from very low confidence (1) to very high confidence (5).

If there were perfect agreement for question 1, all participants would select the same response (100% agreeing on one option). In contrast, perfect disagreement would result in an even split, with 25% selecting each of the four severity options. The study revealed that the average agreement across all cases was only 31%, a result far closer to random chance than to any reliable judgment. Despite this very high level of noise, confidence (question 2) was high, with an average of 4.83 out of 5. Practically, this means that no one agrees, but everyone thinks they are correct. This is a perfect storm for terrible decision-making. Fortunately, energy assessment offers a better approach.

THE "HIGH-ENERGY" STUDY: FROM NOISE TO ANALYSIS

Around the time we were conducting the noise experiment, we were also studying William Haddon's work and experimenting with the energy wheel. Energy theory had proven so effective for hazard recognition, that it got us thinking about logical extensions of energy theory for severity assessment. If energy causes physical harm, maybe more energy causes more harm, and a critical energy threshold could be identified?

Our hypothesis was that the amount of energy present before an incident predicts the severity of the resulting injury. This hypothesis probably makes logical sense as you read it. After all, a heavier object higher off the ground has more potential for serious harm than a lighter object lower to the ground, right? And while a static electricity shock is merely unpleasant, 10 amps is enough to be deadly.

Not only did we think that we could predict the most likely severity using energy assessment, but we also thought it could help us identify the exact amount of energy where a SIF becomes the *most likely* outcome. The stakes demanded a full-scale study, so we collected 505 detailed injury reports from government databases and a consortium of large construction companies.[2] The dataset had approximately equal distribution across the four OSHA severity categories: first aid, medical case, lost time, and fatal/disabling. We used this dataset to test our hypotheses by: (1) estimating the amount of energy for the hazard in each case in Joules[3]; (2) modeling the distribution of energy for each severity level; and (3) computing the respective energy thresholds.

The results, shown graphically in Figure 4.2, supported three conclusions:

1. **Hazards with less than 500 Joules** are most likely to result in first-aid injuries.
2. **Hazards with 500–1500 Joules** are most likely to result in medical cases or lost-time injuries (i.e., typical "recordable" injuries).
3. **Hazards with more than 1500 Joules** are most likely to cause serious injuries or fatalities (SIF).

The most notable result was the 1500-Joule "high-energy" threshold. This is the point where a SIF is not only possible but is the most likely outcome. When a hazard transfers more than 1500 Joules of energy to a human body in one instance, the most common outcome will be a SIF. As the energy increases from 1500 Joules, the likelihood of SIF increases exponentially. What does this mean? 1500 Joules is an objective threshold for determining when a hazard is serious and when it is not. It was through this study that the term "high-energy" was born.

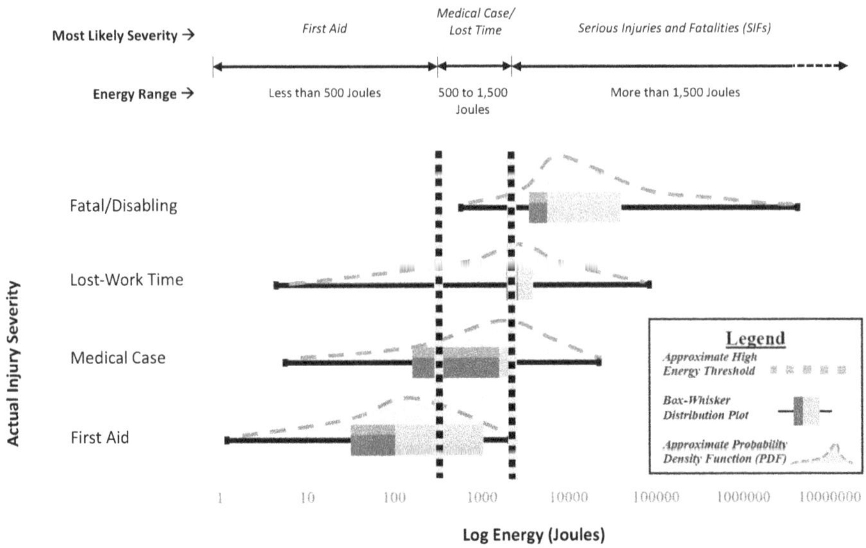

FIGURE 4.2 Distribution of energy by actual severity level. Energy levels less than 500 Joules are most likely to result in a first-aid injury; energy levels between 500 and 1500 Joules are most likely to result in medical case or lost work-time injuries; and energy levels above 1500 Joules are most likely to result in a serious injury or fatality (SIF). These findings were originally published and explained in the referenced journal paper. There are several notable features of this graph: (1) the x-axis is on a log scale so any differences you see in the distributions are *massive*; (2) the distribution of lost-time and medical cases incidents is almost exactly the same (indicating that there are three levels, not four); (3) the 'high-energy' line where a SIF becomes the most likely outcome is approximately 1500 Joules; (4) SIFs are still possible at low energy levels, but they are highly unlikely; and (5) the overlap in distributions means that other factors also matter such as the sharpness of the physical object and where the object strikes the body.

The data for this figure was sourced from Hallowell, M. R., Alexander, D., & Gambatese, J. A. (2017). Energy-based safety risk assessment: Does magnitude and intensity of energy predict injury severity?. *Construction Management and Economics*, 35(1–2), 64–77.

To give you a practical idea of 1500 joules, here are a few examples. This amount of energy is roughly equivalent to lifting 15 kg (33 lbs) to a height of 10 meters (33 feet); running a 1-watt LED bulb for about a half hour; or inflating a bike tire to about 44 psi. Practically speaking, 1500 Joules isn't a lot, and "high-energy" is almost always present in one form or another on a typical worksite.

The high-energy threshold helps us to objectively answer practical questions like, *was the most likely outcome of the near miss a SIF or not?* Or, *which of the hazards that I have identified today have the likely potential to cause serious harm?* Although this high-energy threshold does not guarantee a SIF, it helps us transition from a guess of potential outcome to an objective method of determining SIF potential. It opens the door for a SIF-focused strategy that can be used across the safety system, from early design to incident classification and learning.

The term "high-energy" has a very specific scientific meaning. It's important to be clear: the term "high-energy" has a specific scientific meaning. It's not someone's opinion about how dangerous something looks; it's based on math and physics. Instead of relying on judgment, we can assess severity using real-world measurements like height, weight, speed, and volume, applying basic equations from physics to determine how much energy is present. Although the thresholds could be changed based on better evidence, they are not a matter of opinion.

Not every SIF is caused by high-energy, and not every high-energy hazard will cause a SIF. The high-energy threshold is the point where a SIF is most likely. This does not mean that every SIF involves high energy or that 1500 joules will always cause a SIF. For example, we have all heard stories where someone slipped on the ice and hit their head just the wrong way or cases where someone was able to walk away from a fall from a significant height. These energy assessments do not reflect certainties; they reflect tendencies that are useful for prioritizing in planning, work execution, and learning activities.

CASE EXAMPLE: ESTIMATING ENERGY MAGNITUDE FOR A HAZARD

Appendix 2 includes equations and examples for gravity, motion, electrical, pressure, and mechanical energy. To illustrate how an energy assessment is performed, let's consider the worker on the ladder from Figure 4.1. This is an example of gravity energy related to the worker's position above the ground. Put simply, gravitation energy depends on two factors: height and weight. To be very precise: Gravitational Energy (E) is directly proportional to the mass of an object or individual (m), the height above a reference point (h), and the gravitational constant (G). In the International System of Units (SI), mass is quantified in kilograms (kg), height in meters (m), and the gravitational constant (G) is standardized at 9.8 m/s^2.

Given basic estimates of height and weight, we can compute the amount of energy using the equation for Gravitational Energy, Eq. (4.1).

$$E = m \times h \times G \tag{4.1}$$

For the example in Figure 4.1, let's say the worker's weight is 85 kg (187 lbs) and his height based on his center of gravity is 3 meters (10 ft). The energy magnitude would be computed as follows:

$$E = 85\,kg \times 3\,m \times 9.8\,m/s^2 = 2{,}499\,\text{Joules}$$

Conclusion: High-Energy

Although people might disagree on the exact estimates for height or weight, that's okay because we can always go back, take better measurements, and let the data settle the question. These estimates are grounded in verifiable facts, not gut feel, which makes the conversation more constructive and the conclusions more reliable.

HIGH-ENERGY ICONS

To determine if a hazard is high-energy, you can always do the math. When an incident occurs, you might have the time and resources to do a full energy assessment. However, energy computations are not always feasible for planning and prevention, especially in high volume in the field. For example, it would be ridiculous to ask craft professionals and front-line workers to "do the math" for every hazard they encounter on a worksite. We clearly need an expedited and simplified approach that can be implemented at scale. To this end, we created the high energy icons shown in Table 4.1. These icons represent common hazards that are almost always more than 1500 Joules of energy. For example, the fall from elevation greater than 4 ft icon could be used in lieu of detailed calculations.

TABLE 4.1
Common High-Energy Hazards

Icon	Description
 Suspended Load	Most suspended loads require specialty equipment to lift massive objects off the ground. In such a case, suspended loads are over the high-energy threshold. Suspended loads typically have gravitational energy from the vertical height and kinetic energy from the lateral movement. These two energy sources should be treated separately as they require different types of protection.
 Fall from Elevation	Considering the average weight of a human is approximately 200 lbs (90 kg), 4 feet (1.22 m) measured from the ground surface to the bottom of the feet exceeds the high-energy threshold. The approximate center of gravity at this elevation is roughly 7 feet (2 m), which is over the 1500-Joule threshold.
 Mobile Equipment/Traffic with Workers on Foot	For this icon, the energy exposure is taken from the point of view of the worker on foot and not the equipment operator or vehicle driver. Because of the mass, most mobile equipment, including motor vehicles, exceeds the high-energy threshold when the equipment or vehicle is in motion. *Note*: For work zone traffic only, an incident occurs only when a vehicle departs from the intended path of travel and is within 6 ft of an exposed employee or if an employee enters the traffic pattern.
 Motor Vehicle Occupant	Estimations of the motor vehicle speed typically involved in serious or fatal crashes vary greatly from the National Transportation Safety Board, National Highway Transportation Safety Association, and the US Department of Transportation. The team selected a conservative estimate of 30 miles per hour as the high-energy threshold. This energy exposure is taken from the point of view of the vehicle occupants including the driver, but not workers on foot.

(*Continued*)

TABLE 4.1
(Continued)

Icon	Description
 Heavy Rotating Equipment	Computing mechanical energy can be complex. Therefore, all heavy rotating equipment beyond powered hand tools typically exceed the high-energy threshold and any rotating equipment or tools exceeding 100 rotations per minute (rpm) should be considered high- energy.
 High Temperature	Solid surfaces over 150°F are considered high-energy, if contact is possible with a surface area of 100 square inches or larger. Per the American Burn Association,[a] contact with any surface at or above 150°F can cause third-degree burns within at least 2 seconds, reflecting typical human reaction time. The 100 square inches correspond to approximately 5% of the body surface area of a fifth percentile female.
 Steam	According to the American Burn Association, industrial exposures to steam almost always exceed the high-energy threshold.
 Fire with Sustained Fuel Source	According to the North American Combustion Handbook,[b] a lightly combustible material like paper burns at approximately 700 degrees Fahrenheit, far exceeding the temperature threshold. Fire with a sustained source of fuel exceeds the high-energy threshold.
 Explosion	Most incidents described as an explosion exceed the high-energy threshold.
 Excavation or Trench	An exposure to unsupported soil in a trench or excavation that exceeds 5 feet of depth (1.5 m) exceeds the high-energy threshold. Typically, for each foot of depth, soil pressure increases by about 40 pounds (2 kPa) per square foot. Thus, at 5 ft (1.5 m) of depth the pressure is approximately 200 psf (10 kPa).

(Continued)

TABLE 4.1
(Continued)

Icon	Description
 Electrical Contact with Source	According to NFPA 70E,[c] electricity equal to or exceeding 50 volts typically produces current sufficient to result in serious injury or death.
 Arc Flash	Arc flashes typically exceed the high-energy threshold because of the voltage exposure according to the NFPA 70E.
 High Dose of Toxic Chemical or Radiation	Exposure to toxic chemicals, radiation, or biological agents. An industrial hygienist, chemist, toxicologist, or other competent person should be involved in the assessment of toxicity and acceptable exposure limits. The following references should be used to judge acceptable exposure limits: • Immediately Dangerous to Life or Health (IDLH) Values from the Center for Disease Control[d] • Exposures which reduce oxygen (O_2) levels below 16% • Corrosive chemical exposures (pH <2 or >12.5)
 Energy Calculations	Some hazards require full energy assessment if they do not fall into one or more of the categories above. To determine whether a hazard exceeds 1500 Joules or not, the energy magnitude must be estimated based on physical factors using the equations in Appendix 2.

Hallowell, M. R., & Spencer, C. (2024). Safety classification & learning model: Defining & classifying potential serious injuries & fatalities. *Professional Safety*, 69(01), 18–26

[a] American Burn Association, Scald Educator's Guide. Available at: https://ameriburn.org/wp-content/uploads/2017/04/scaldinjuryeducatorsguide.pdf

[b] Reed, Richard, J. (1978). *North American Combustion Handbook: A Basic Reference on the Art and Science of Industrial Heating with Gaseous and Liquid Fuels*. North American Mfg. Company, Philadelphia, PA

[c] National Fire Protection Association, Standard for Electrical Safety in the Workplace (NFPA 70E). Available at: https://www.nfpa.org/codes-and-standards/all-codes-and-standards/list-of-codes-and-standards/detail?code=70E

[d] Center for Disease Control, Immediately Dangerous To Life or Health (IDLH) Values. Available at: https://www.cdc.gov/niosh/idlh/intridl4.html

USING THE HIGH-ENERGY ICONS

The high-energy icons serve as a practical tool for identifying high-energy hazards. Designed to enhance consistency and collaboration, these icons make it easier to determine which hazards are high-energy. Below, we will explore their purpose, scope, and governance.

1. **The purpose of the icons is to provide examples of the most common hazards that are high-energy**. These icons represent hazards that are, for all practical purposes, *categorically* over 1500 Joules. For example, heavy mobile equipment exceeds 1500 Joules any time it is moving at a discernable speed. We have found that those included in Table 4.1 represent approximately 85% of all high-energy hazards. Any common hazard that is always over the 1500-Joule threshold is a candidate for an icon.
2. **This is *not* intended to be a comprehensive list of hazards that could cause SIFs**. There are many hazards that are *sometimes* high-energy such as tools at height, confined spaces, hoses under pressure, tension in cables, and power tools. However, these are not included as high-energy icons because they are not always over 1500 Joules. For these, we must do the math using the equations in Appendix 2. Take dropped objects for example. A 1-lb screwdriver at 5 feet of height is not high-energy, but a 5-lb wrench at 300 feet is high energy. The conclusion for dropped tools depends on the specific factors.
3. **If applicable, defer to a conservative government standard**. Some government agencies maintain thresholds for the "seriousness" of hazards. For example, OSHA sets the threshold for electrical hazards at 50 volts and the National Transportation Safety Board (NTSB) uses 30 mph for motor vehicle occupants. For these icons, we defer to the government standards if they are conservative (i.e., less than 1500 Joules). These may change depending on geography and governing body.
4. **These icons must be governed**. To promote consistency in application and enable collaboration, the high-energy icons should be governed by communities of practice. For example, in the formative years of EBS, the Edison Electric Institute (EEI) provided governance for the electric utility sector. When a change is suggested, the community still considers, debates, and eventually votes on the proposal. Traditionally, companies would have customized, adapted, and added to the original icons based on their preferences. Before long, this would have resulted in countless high-energy icons and a complete lack of consistency. This new approach to governance is an excellent example of compromise for the sake of collaboration. Although the high-energy icons are applicable to almost any industrialized work, new sectors may wish to establish communities of practice to create and govern their own icons. For now, the EEI community has set and maintains a standard, which we have applied in our research across all industry sectors and around the world without need for modification.

5. **Maintaining consistency is not easy**. In 2021, the EEI community of practice considered changing the electrical contact icon from 50 volts to 480 volts. Some felt that we should defer to the conservative government standard propagated by OSHA, while others believed that we should prioritize the 1500-Joule threshold.[4] The group deliberated for a month and voted. In a 51% to 49% vote, the majority opinion was to maintain the 50-volt threshold. Contrary to our previous tendencies in the safety profession, the 49% who were outvoted agreed to support the decision and move on in the name of collaboration. This is a big win for SIF prevention and a mark of progress.

THE STUFF THAT KILLS YOU (STKY)

Safety is about protecting the men and women in the field. To be successful, front-line employees must understand and genuinely embrace the concept of high-energy. Unfortunately, the terms “energy” and “joules” and even the icons can be perceived as academic nonsense. This isn’t the language of the field, which is an understandable barrier to implementation.

In a simple but genius move, some companies have adjusted the messaging into language that resonates much better at the front line, coining terms like the Stuff that Kills You (STKY).[5]

It is immediately obvious that STKY is useful. Rather than say, let’s *identify the high-energy hazards with over 1500 Joules*, we can ask our front-line workers to *identify the stuff that kills you* or, even better, *identify the STKY*. The language around STKY is easy to remember, fun to use, and puts attention on what matters most (Figure 4.3). In the words of a senior welder in one of my training sessions after hearing about STKY, “you safety guys finally get it.”

If used the right way, STKY can simplify the messaging *and* carry research rigor. A common pitfall is to encourage teams to use their judgment when identifying STKY. What we *think* is STKY is not relevant when there are methods to *know* when something is STKY and when it is not. When relying on judgment alone, crews often classify many low-energy hazards as STKY because they consider the worst possible outcome, which is always a SIF. Furthermore, because energy assessment can be complex and counterintuitive, crews may also undervalue some high-energy hazards, not realizing they are actually STKY. For instance, the potential impact of hazards like high surface temperatures and pressurized tanks is frequently underestimated

I use the terms high-energy and STKY interchangeably because, to me, they have the same meaning. If I am speaking with an executive, engineer, or architect, I might say high-energy and give examples in Joules. However, if I am speaking with a work crew or a construction manager, I am more likely to use the term STKY and share the high-energy icons as examples. If we take STKY to mean high-energy and not something else, we can be more flexible with our language while preserving the research rigor behind the concept. This allows us to choose language that is tailored to our audience.

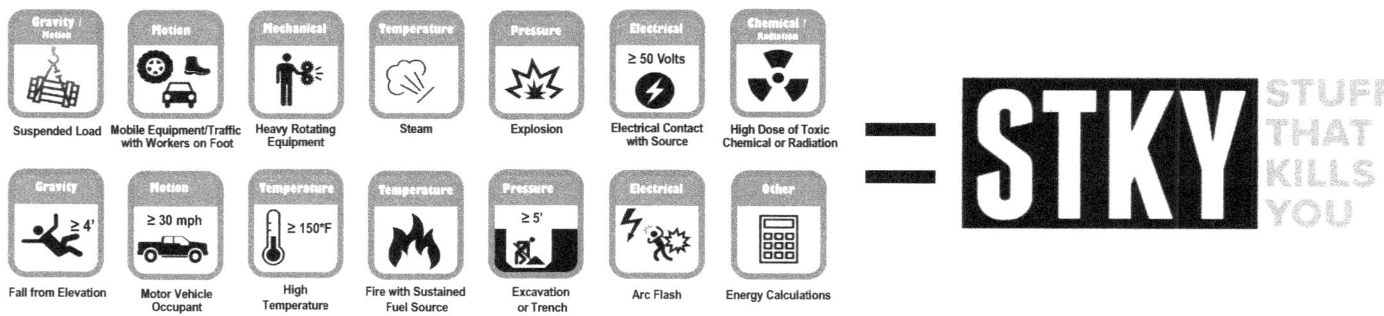

FIGURE 4.3 STKY is a field-friendly shorthand for high-energy hazards. The term STKY is used in the field to describe hazards with the potential to cause serious harm. This concept maps directly to high-energy hazards, typically those exceeding 1500 Joules.

RELATIONSHIP BETWEEN THE ENERGY WHEEL AND HIGH-ENERGY ICONS

The relationship between the energy wheel and high-energy icons may be the most confusing part of EBS. Since both relate to hazards and are visual tools, it can raise questions: Are they the same thing? Should we replace the energy wheel with the high-energy icons? Is one a subset of the other?

The energy wheel and high-energy icons (STKY) are distinct but intended to be used together. While the energy wheel is a tool to see all types of hazards, the high-energy icons are a tool to identify which have the likely potential to cause SIFs. As an analogy, the energy wheel opens the aperture to capture everything, and the high-energy icons close the aperture to focus on what matters most.

To use the energy wheel and high-energy icons together, follow these two steps:

Step 1: Use the energy wheel to see as many hazards as possible.
then…
Step 2: Use the high-energy icons to identify which of those hazards are STKY.

As companies mature and increase their "energy-IQ," they may find more efficient ways to use the energy wheel and high-energy icons in their safety practices.

CASE EXAMPLE: IDENTIFYING HIGH-ENERGY HAZARDS IN A WORK TASK

Let's practice the first two steps of EBS using the image in Figure 4.4. Our goal is to identify all relevant energy sources and the subset of these hazards that are high-energy. In the image provided, we have a pipeline crew installing shacks for a subsequent welding operation. I have selected this image because it has many hazards and because we have analyzed it deeply in the research. As you review the image, you can make assumptions as needed for the sake of this paper exercise, but don't get too caught up in the details of a picture. In real life, we would be able to test our assumptions by asking questions in the field.

For this activity, complete the following steps:

1. Use your instinct and experience to identify as many hazards as you can and write them down. Remember, missing personal protective equipment (PPE) and factors like paying attention are not considered hazards because they are not sources of energy.
2. Use the energy wheel to find hazards that you may have missed and add them to your list.
3. For every hazard on your list, use the icons or energy calculations to assess whether it is high energy (STKY) or not. Put a star next to the high-energy hazards. They will get special treatment as described in the next chapter.

In my assessment, I found a total of 15 hazards. For each hazard, I performed an energy assessment as summarized in Figure 4.5. You can use this figure to check your work but please be aware that this is not exhaustive, and our assessments may differ if we made different assumptions. The purpose of the exercise is to provide a visual example that accompanies the written description.

FIGURE 4.4 Pipe shack installation case.

This figure appeared in: Erkal, E. D. O., & Hallowell, M. (2023). Moving beyond TRIR: Measuring & monitoring safety performance with high-energy control assessments. *Professional safety*, 68(05), 26–35. Reprinted with permission.

Hazard ID	Hazard Name	Energy Source	High-Energy?	Assessment Reasoning
H1	Suspended load (falling)	Gravity	Yes	Suspended Load Icon
H2	Elevated pipe	Gravity	Yes	Suspended Load Icon
H3	Side boom (tipping)	Gravity	Yes	Heavy Mobile Equipment/Traffic Icon
H4	Side boom (tracking)	Motion	Yes	Heavy Mobile Equipment/Traffic Icon
H5	Swinging load	Motion	Yes	Suspended Load Icon
H6	Vehicular traffic	Motion	Yes	Heavy Mobile Equipment/Traffic Icon
H7	Cable and pulley	Mechanical	Yes	Tension Calculations
H8	Power lines	Electrical	Yes	High Voltage and Arc Flash Icons
H9	Compressed gas	Pressure	Yes	Pressure Calculation
L1	Uneven ground	Gravity	No	Gravity Calculation
L2	Construction noise	Sound	No	Sound Calculation
L3	Sun exposure	Radiation	No	Under Permissible Exposure Limit
L4	Insects	Biological	No	Biological Toxicity Assessment
L5	Pesticides	Chemical	No	Chemical Toxicity Assessment
L6	Pipe Surface Temp.	Temperature	No	Temperature Calculation

FIGURE 4.5 Pipe shack installation case hazard identification results . This assessment includes all energy sources identified by the observer. Hazards with calculated energy levels below 1500 Joules were classified as low energy.

This figure appeared in: Erkal, E. D. O., & Hallowell, M. (2023). Moving beyond TRIR: Measuring & monitoring safety performance with high-energy control assessments. ***Professional safety*****, 68(05), 26–35. Reprinted with permission.**

CONCLUSION

Imagine for a moment that you are on an active worksite, just before work is about to commence. Using experience and the energy wheel as a guide, you have identified as many hazards as possible. You have considered the task and the environment, and you are confident that your scan is complete. Now, it is time to identify which are high-energy (STKY) and which are not.

To do this, go one by one through the full list of hazards. Use the icons in Table 4.1 to identify the high-energy hazards. This will save you time. For any hazards that are uncertain or where an icon doesn't apply, use the energy calculator. With this assessment, you should have a full list of hazards and a designation next to those that are high-energy (STKY). Although all hazards should have a countermeasure, high-energy hazards require special controls that minimize the risk of a SIF. This will be the subject of our next chapter.

There you have it: The start to a SIF-focused strategy. An energy assessment method that is based on physics, medicine, and backed by research. Since energy assessment is fundamental, it will transcend industry sectors, geographies, and cultures. Whether you are in your office or on the moon, energy is energy and people are people. This method requires no customization.

Remember, energy is not our enemy. We need it to get the job done. Safety is about harnessing this energy to do our work, without letting that energy contact a person. When the amount of energy is high enough, more than 1500 Joules to be exact, it is life threatening. For these high-energy hazards, we must have robust and resilient controls in place because the lives of our workforce depend on it. That is the subject of our next chapter.

KEY TAKEAWAYS

1. The amount of energy associated with a hazard predicts the severity of the injury. More energy causes more harm.
2. A SIF is the most likely outcome when a hazard has more than 1500 Joules of energy. These are known as "high-energy" hazards.
3. The term stuff that kills you (STKY) was popularized as a plain language way to describe hazards that can cause SIF. The terms "high-energy" and "STKY" are synonymous.
4. Using energy assessment to identify what is STKY instead of our judgement improves consistency, increases focus, and enables collaboration. It also helps us to transition from certainty (worst possible outcome) to tendency (most likely outcome).
5. The focus on high-energy helps organizations to assign priority and resources to the hazards most likely to cause SIFs.

NOTES

1 A noise experiment is conducted by independently providing subjects with the same information and measuring the consistency in their assessments. For more information about this type of research, I highly recommend the book *Noise* by Nobel Prize-winning author Daniel Kahneman.

2 Hallowell, M. R., Alexander, D., & Gambatese, J. A. (2017). Energy-based safety risk assessment: Does magnitude and intensity of energy predict injury severity?. *Construction Management and Economics*, 35(1, 2), 64–77.

3 A Joule is the metric unit for energy, which indicates the amount of work that can be completed with that energy. In Imperial units, energy is measured as foot-pounds (ft-lb) and 1 Joule is equivalent to 0.74 ft-lbs.

4 The 1500 Joules is roughly equivalent to 1 amp of current. With typical resistance of the human body, 1 amp is roughly equivalent to 480 volts with typical human resistance.

5 Pronounced "sticky.".

5 Direct and Alternative Controls

Principle: Controls save lives
Practice: Control the Stuff that Kills You (STKY) with Direct and Alternative Controls

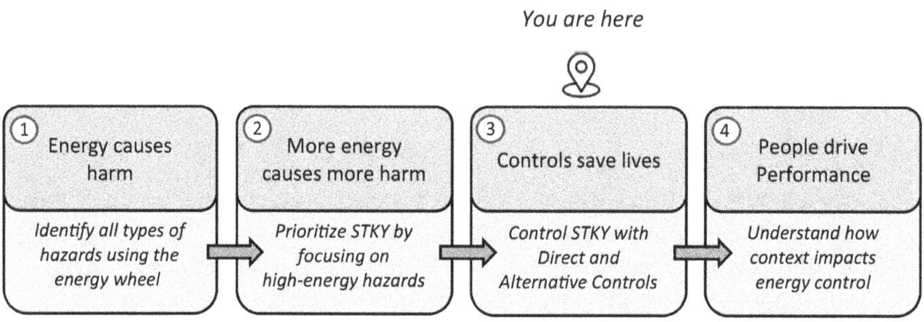

SIF prevention hinges on one essential factor: effectively controlling high-energy hazards. The critical question is, what level of control is "enough"? For example, if we are working around heavy rotating equipment, is situational awareness sufficient? How about caution tape and a sign? A machine guard? Or do we need to completely de-energize the system? While regulations provide some prescriptive guidance, clear and generalized rules help us consistently evaluate control options and determine if they are adequate to prevent SIFs.

Foundational tools and frameworks provide valuable ways to classify and analyze controls. For example, Haddon's 10 countermeasures provide a systematic approach to mitigating risk such as eliminating energy, reducing energy magnitude, or interposing physical barriers.[1] Similarly, frameworks like the hierarchy of controls[2] and analytical tools such as bowtie models[3] help guide decision-making.

These models are valuable, but they often do not tell us when a level of protection is sufficient and when it is not. While the complete elimination of hazardous energy clearly provides the most robust protection, it is often impractical. In such cases, other measures like barriers and personal protective equipment (PPE) play essential roles. This chapter moves from categorizing and describing controls to a rule-based approach for determining whether controls for high-energy hazards are sufficient.

DOI: 10.1201/9781003641483-6

INTRODUCING THE "SAFETY ONION"

At its essence, safety is about controlling energy, with SIF prevention requiring a focus on high-energy hazards. Everything else we do in safety, whether it involves culture, leadership, or trust, exists to support this fundamental objective. Some may view this perspective as downplaying these other elements, but the reality is quite the opposite. They are not separate from energy control; they are what make it possible.

Many organizations focus heavily on the outer layers of what I describe in this book as the "safety onion," a new framework introduced here to illustrate the layers of safety effort (see Figure 5.1). These layers emphasize relationships and behavior but can often be disconnected from the primary goal of controlling energy. While these elements are important, prioritizing them without anchoring efforts in high-energy controls creates an imbalance. Initiatives such as leadership engagements, climate surveys, and stand-downs may enhance visible aspects of safety culture, but if these activities fail to target high-energy hazards, we risk contributing to the SIF plateau.

The "safety onion" illustrates the relationship between high-energy controls and their supporting elements. At the center are tangible measures to mitigate or neutralize hazardous energy, surrounded by layers of tactics, strategy, and culture that enable and sustain these controls. The closer a layer is to the center, the more directly it is connected to the risk, while the outer layers shape organizational priorities and behaviors that enable energy control. For instance, fostering psychological safety and trust helps organizations identify gaps in controls and allocate resources effectively.[4] But no matter how strong the outer layers are, they cannot compensate for weaknesses at the core. When all layers align, they create a system capable of breaking the SIF plateau.

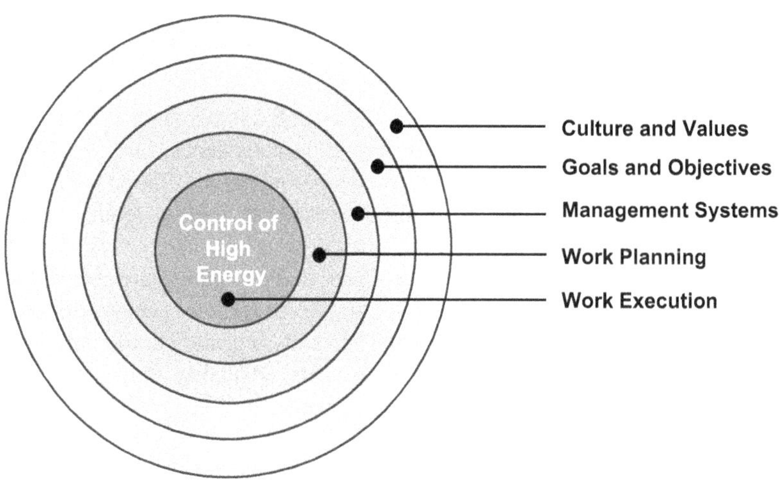

FIGURE 5.1 Layers of the SIF prevention onion. At the core are tangible controls that directly mitigate high-energy hazards, surrounded by enabling layers such as planning, systems, leadership, and culture. Each layer supports the presence and effectiveness of energy controls in the field.

REDEFINING SAFETY AS THE PRESENCE OF CONTROLS

Focusing on energy control requires an updated definition of safety. Popularized by Human and Organizational Performance (HOP), safety has been reimagined as "the presence of safeguards, not the absence of injuries." This perspective challenges longstanding beliefs about what it means to work safely and shifts attention from injury-free outcomes to the systems that actively manage risk. It recognizes that day-to-day results, including the absence of injuries, are often poor indicators of true safety.

To illustrate this concept, consider the following scenario:

> I am driving my car from Boulder to Denver in the middle of January. Heavy snow is falling, and I am speeding 10 mph over the limit. I am driving a Honda Civic with bald tires and wiper blades that should have been replaced years ago. The radio is blaring, my kids are arguing in the back, and I am steering with my knee while eating a burrito.

Even if we arrived in Denver without incident, this was not a safe drive. The absence of an incident would not make my actions safe. On the other hand, imagine I took every precaution on the drive. I could have a four-wheel-drive car with new tires, full focus on the road, and controlled speed, but still be sideswiped by an erratic driver. The outcome alone does not describe the safety of the situation.

This story highlights that safety should not be measured by outcomes alone. Safety is best described by the level of control we have over hazardous energy. With this shift in mindset comes the need to answer critical questions: What is a control, and when is it sufficient? Answering these questions is essential for putting this new mindset into practice.

WHAT IS AN ENERGY CONTROL?

As we turn our attention to energy control, it's essential to distinguish what qualifies as a control and what does not. Not everything casually labeled a "control" effectively manages the boundary between high-energy and people. To reduce confusion, we define "energy controls" as physical measures that actively reduce or remove hazardous energy.

Many safety measures, such as policies, plans, and procedures are critical for supporting energy control, but they do not mitigate hazardous energy directly. For instance, safety briefs and permits are vital for ensuring that controls are properly installed but they are not energy controls themselves. To qualify as an energy control, a measure must meet the following criteria, known as the 3Ts[5]:

1. ***Timely*****:** The control must be actively in place and functioning during the work period when the high-energy hazard is present. This ensures the control is in use exactly when the risk exists.
2. ***Tangible*****:** The control must be physically present and observable on the worksite. This means that it must be something that workers can see, interact with, or verify, rather than paper-based or administrative measure.

3. ***Targeted***: The control must be specifically designed and installed to address the identified high-energy hazard. This ensures the control directly mitigates the risk associated with the energy source, rather than being a general or unrelated safety measure.

Some examples of safeguards that meet the 3T criteria include guardrails, de-energization with lock-out tag-out, proximity detection systems, spotters, and many forms of specialty PPE. In contrast, measures like rules, procedures, situational awareness, and training, while vital to safety programs, do not qualify as energy controls because they fail to meet one or more of the 3Ts. Rather than being classified as energy controls, these elements are more accurately positioned in the adjacent layers of the safety onion, where they support but do not directly control hazardous energy.

To further illustrate the distinction between energy controls and supporting safety measures, a variety of common safety measures are evaluated against the 3T criteria in Table 5.1. This comparison highlights which measures qualify as energy controls and which do not. For example, although pre-job briefs are incredibly important, they do not qualify as an energy control because they are not present during work, are not tangible, and are not dedicated to any one hazard.

The examples provided are illustrative and meant to demonstrate how the criteria for Timely, Tangible, and Targeted can be applied. The specific classification of a safety measure may vary depending on the context in which it is used. For example,

TABLE 5.1
Assessment of Potential Controls for the Case Example

Safety Measure	Timely?	Tangible?	Targeted?	Is It an Energy Control?
Spotter for an excavator	✓	✓	✓	Yes
Proximity detection system	✓	✓	✓	Yes
Cones with caution tape	✓	✓	✓	Yes
Temporary machine guarding	✓	✓	✓	Yes
Arc flash PPE	✓	✓	✓	Yes
Fall protection system	✓	✓	✓	Yes
Load limiting devices on cranes	✓	✓	✓	Yes
De-energize and lock-out tag-out	✓	✓	✓	Yes
Permanent guardrail at height	✓	✓	✓	Yes
Training and experience	✓	X	✓	No
Situational awareness	✓	X	X	No
Psychological safety	✓	X	X	No
Safety slogan	✓	X	X	No
Safety walks	✓	X	X	No
Emergency response plans	X	X	✓	No
Risk assessment	X	X	X	No
Standard operating procedures	X	X	✓	No
Pre-job safety brief	X	X	X	No

cones with caution tape may be considered Targeted if explicitly designed to address a high-energy hazard, but they may fail this criterion if used generically.

Defining energy controls using the 3T criteria is an important starting point because it distinguishes controls from administrative elements, but it does not address the critical question of whether the control is sufficient to prevent SIFs. That requires further assessment.

In the EBS paradigm, we have defined two types of energy controls: Direct and Alternative Controls. While Direct Controls address the energy at its source, Alternative Controls mitigate the potential for unintentional human error when a Direct Control is not feasible. Together, these categories help us to determine whether controls are sufficient to protect against high-energy, serving as the cornerstone of SIF prevention.

DIRECT CONTROLS: THE GOLD STANDARD OF HIGH-ENERGY PROTECTION

Direct Controls are a subset of energy controls that set the highest standard for SIF prevention by eliminating, reducing, or isolating high-energy at its source. Like the 3Ts, we established rule-based criteria for Direct Controls to support consistency and collaboration.[6] Based on the collective wisdom of our community, these rules clarify the features of an energy control that make it sufficient.

Direct Control Rules

As a specialized subset of energy controls, a Direct Control must meet two additional criteria *in addition to the 3Ts*: it must effectively mitigate the high-energy hazard and remain reliable even if someone makes a mistake. Described in detail below, these rules are designed to be strictly applied.

1. ***Mitigates High-Energy:*** *The control must effectively mitigate exposure to the high-energy hazard when installed, verified, and used properly.*

   A Direct Control must address hazardous energy through one of the three methods: elimination, reduction, or isolation. These methods directly target the energy source and separate the high-energy required for work from the people in its vicinity.

   - *Elimination* is when an energy source or the worker are completely removed from the hazardous environment, removing any possibility of exposure to high-energy. This control requires a definitive and verifiable action to ensure that no residual energy remains. Elimination can be achieved by removing the energy source from the worksite entirely, or by installing permanent or semipermanent structures such as guardrails that fully neutralize the energy. Example: Depressurizing a gas line, verifying with a pressure gauge that it has been fully neutralized, and isolating it using lock-out tag-out to ensure no pressure is reintroduced.
   - *Reduction* is when the energy magnitude is reduced from a high-energy state to a low-energy state, meaning the energy is brought below the threshold of

1500 Joules where a SIF is no longer the most likely outcome. Example: Fall arrest systems dissipate the kinetic energy of a falling worker to less than 1500 Joules, reducing the force exerted on the body to a survivable level.

- *Isolation*: Energy is not reduced to a low-energy state, but employees are shielded from energy exposure by a temporary physical barrier designed to block the hazard under normal operating conditions. These barriers must be properly maintained, tested, and free of damage to function as an effective Direct Control. Often, they are referred to as specialty PPE. Example: Arc-rated flash suits with full coverage worn during electrical switching provide a physical barrier that protects workers from thermal energy in the event of an arc flash.

DOES IT MITIGATE ENERGY?

✓ Installing a trench box for pipeline work in a deep trench. *Yes, this mitigates energy by isolating the workers from the soil pressure from the trench.*

✖ The company had a safety stand-down to discuss trench safety. *No, the stand-down itself does not mitigate energy.*

2. **Works even when someone makes a mistake:** The control must be effective even if there is unintentional human error during the work that is unrelated to the installation of the control.

The field of psychology has taught us that error is an inherent part of the human condition. HOP scholars encapsulate this idea with a simple and powerful statement: "Even the best people make mistakes.[7]" Regardless of how mindful, well-trained, or careful someone is, they will inevitably make a mistake (it's not if they will make a mistake, it's when). Therefore, we should not be surprised when errors occur. Instead, we should accept that errors are inevitable and plan accordingly.

With this understanding and a desire to align with HOP, a Direct Control must work even when someone makes a mistake. That is, a control must be robust enough that workers can be momentarily distracted, forget a step, or stumble, and not die as a result.

Functionally, this means that any control that *is* a person or is actively *operated by* a person does not satisfy this rule. Controls such as flaggers, spotters, and operators require human perfection over extended periods, which we know to be impossible. The same is true for safety measures like fire extinguishers and tag lines that must be operated by a person who could make a mistake in the heat of the moment.

Controls that remain effective in the presence of human error are physical or technological in nature. For example, a hard physical barrier near a leading edge will work even if someone trips and an automatic fire suppression system will function independent of an operator. These types of barriers only fail if they are improperly designed, poorly installed, or if someone intentionally defeats them. While any control can be intentionally bypassed, our concern is with ensuring that controls remain effective in the face of unintentional error.

Interestingly, many controls that satisfy this rule exist specifically because we expect errors. For instance, we install fall protection systems assuming someone will eventually stumble or drift too close to an otherwise unprotected edge.

> **DOES IT WORK EVEN IF SOMEONE MAKES A MISTAKE?**
>
> ✓ Concrete barrier installed between heavy equipment and a walking path. *Yes, the control works even if a worker on foot or the equipment operator makes a mistake.*
>
> ✖ A highly skilled heavy equipment operator uses their cameras to monitor workers on foot. *No, the equipment operator could lose focus and strike a worker.*

ACTIVITY: IS IT A DIRECT CONTROL?

The following activity provides a practical assessment of various safety measures to determine if they meet the criteria for Direct Controls. To guide this evaluation, we assume the following:

- All measures are installed, verified, and used properly.
- The workforce is well-trained, and appropriate enabling activities, such as pre-job safety meetings, have taken place.
- The control meets the 3T criteria.

Table 5.2 evaluates the same safety measures included in Table 5.1, now applying the two additional Direct Control rules. All safety measures that did not meet the 3T criteria in Table 5.1 have been removed. This activity illustrates how to distinguish Direct Controls from other energy controls like spotters, proximity detection systems, and cones that do not meet the criteria.

LESSONS LEARNED ABOUT DIRECT CONTROLS

In the years since we defined Direct Controls, we have learned a great deal about putting them into practice. Along the way, we've encountered key challenges, such as the limitations of even the best-designed controls and the need for robust systems to ensure their success. We have also designed and tested Alternative Controls, which are a vital interim solution for situations when Direct Controls are not yet feasible.

TABLE 5.2
Assessment of potential Direct Controls

Safety Measure	Timely Tangible and Targeted?	Mitigates Energy?	Works if someone makes a mistake?	Direct Control?
Temporary machine guarding	✓	✓	✓	Yes
Arc flash PPE	✓	✓	✓	Yes
Fall protection system	✓	✓	✓	Yes
Load limiting devices on cranes	✓	✓	✓	Yes
De-energize and lock-out tag-out	✓	✓	✓	Yes
Permanent guardrail at height	✓	✓	✓	Yes
Spotter for an excavator	✓	X	X	No
Proximity detection system	✓	X	X	No
Cones with caution tape	✓	X	X	No

Always Consider the Details

The activity in Table 5.2 demonstrates the application of rules to evaluate Direct Controls. However, it is always important to consider the full context to ensure accurate and consistent conclusions. For instance, whether a control is vulnerable to human error often depends on the specific nature of the control and its application. Take the example of a crane operator moving a suspended load: the crane operator may be susceptible to error, but if the equipment includes advanced features such as an automated brake, a load-limiting device, and load stabilization, the system may meet the criteria for a Direct Control. Ultimately, the effectiveness of a control depends more on the details than on the name that we assigned to the control. While reference lists of Direct Controls can be valuable, we must always consider the details of each individual situation and apply the rules accordingly.

Every Control Is Vulnerable to Mistakes

Direct Controls are designed to remain effective even when someone makes a mistake during work, but mistakes during design, installation, or verification can still render them ineffective. For instance, a hard physical barrier may meet all Direct Control criteria, yet errors such as engineering miscalculations, incomplete assembly, or missed inspections could compromise its functionality. These vulnerabilities underscore the need for rigorous oversight during all phases of control implementation.

We will explore these human factors further in this chapter, focusing on how to identify and address common challenges that we faced during the design, installation, and verification of Direct Controls.

Every Component of a Direct Control Must Function Effectively

Most Direct Controls rely on a system of elements working together to mitigate hazardous energy. For example, fall protection systems, lifting and rigging methods, and excavation support depend on multiple components operating in unison. To qualify as a Direct Control, all energy-mitigating components must be present and perform as intended. A failure in even one element can undermine the effectiveness of the entire system. Therefore, in practice, we consider a Direct Control to be absent if there is a deficiency in any functional element in the system.

Not Every High-Energy Hazard Has a Direct Control

Ideally, we would have a Direct Control for every high-energy hazard, but this is not always feasible. Our research shows that even among industry-leading companies, only 65 to 75% of high-energy hazards have Direct Controls that meet the defined criteria.[8] In dynamic environments such as construction, mining, and forestry, many high-energy hazards lack feasible Direct Controls, not because companies fail to implement them, but because no viable solutions currently exist. That is, society *as a whole* does not have a feasible Direct Control for many work scenarios. We call these “STKY Situations.[9]”

Examples of pervasive STKY Situations include work zone traffic control, workers on foot near heavy equipment, lateral movement of suspended loads, and troubleshooting on energized equipment. These challenges are not unique to any one company or sector and highlight an industry-wide need for innovation.

When Direct Controls are not feasible, Alternative Controls serve as a temporary yet vital solution. Alternative Controls bridge the gap, focusing on minimizing human error while we work toward long-term Direct Control solutions.

ALTERNATIVE CONTROLS: MINIMIZING POTENTIAL FOR ERROR WHEN ENERGY CANNOT BE MITIGATED DIRECTLY

When a Direct Control is deemed unfeasible and this determination is confirmed by a competent person, the focus must shift to Alternative Controls. When a high-energy hazard is present and a Direct Control is not feasible, workers are left physically exposed and must rely on their knowledge, awareness, and instinct to stay safe. Some of my colleagues have referred to this as a “naked exposure” because there is no room for error. In such a situation, we either must cease work or ensure that Alternative Controls are present to minimize the risk of a critical mistake.

Consider the following example:

> You are a field supervisor, and one of your crew leads calls you. They inform you that workers on foot must be close to an excavator to manually monitor and adjust a piping system with a tight tolerance. The work is urgent, and the crew believes that no Direct Control is feasible.
>
> Your first instinct as a supervisor is to gather more details about the situation. Assuming the crew is correct, and a Direct Control is not feasible, the next logical question is, “What controls do you have in place?”

Responses to this question can vary widely, ranging from robust layers of control to inadequate ones like, "We will keep our head on a swivel." With lives at stake, we need a consistent way to determine when Alternative Controls are sufficient, and when they are not.

Alternative Control Systems

Because Alternative Controls focus on minimizing human error, they are never standalone solutions. Thus, to qualify as a sufficient Alternative Control, the system must include at least two complementary and independent methods of error reduction, and each of these measures must meet the 3T criteria. Multiple elements are essential because no single measure alone can reliably address the complexity of human error.[10]

Psychologists have found that the best way to reduce human error is to use several different methods at the same time, each one helping in a different way. When each element operates independently and serves a distinct purpose, the probability of all elements failing simultaneously is the product of their individual failure probabilities.[11] Complementary controls are not merely additive, they are multiplicative. This is why more than one element is necessary, and they must serve different functions.

To make this concept practical, we categorize Alternative Controls based on the specific functions they serve in reducing human error. When selecting multiple elements, the controls must come from more than one of the following categories:

- ***Physical Obstacle*:** An obstruction that hinders progress toward a high-energy hazard (e.g., non-engineered barricade like cones with toppers).
- ***Dedicated Monitoring*:** Devoted and continuous attention to a high-energy hazard (e.g., a spotter or proximity detection system).
- ***Visual Reminder*:** A visible warning of the presence of a high-energy hazard (e.g., specialty signage).

Consider a work scenario where an excavator is traversing a site with workers on foot. Due to financial and logistical constraints, a Direct Control like a concrete barrier is not feasible. To reduce the likelihood of error, the company implements a system of Alternative Controls: a spotter and an exclusion zone marked with cones and caution tape. These controls are independent and serve distinct, complementary functions. The spotter maintains communication with the operator, while the exclusion zone visually reminds workers on foot to stay at a safe distance.

While exact error rates are rarely known with certainty, illustrative examples can help convey the value of complementary controls. Assuming the error rate of the spotter is 1/1,000 instances per day and the rate of workers entering a marked exclusion zone is also 1/1,000 instances per day, the probability that they will fail concurrently becomes the product: one in a million ($1/1{,}000 \times 1/1{,}000 = 1/1{,}000{,}000$).[12] Although never zero, multiple complementary controls can reduce the probability of error to minimal levels, making "naked exposures" a little less daunting.

If two controls are redundant instead of complementary, the error rate would be closer to the sum of the rates than the product. For example, the error rate with two signs marking an exclusion zone would be 1/2,000 instances (1/1,000 + 1/1,000 = 1/2,000). This is not exact science, but it illustrates how multiple, independent, and complementary controls are desirable for error minimization.

THE HIERARCHY OF ENERGY CONTROL (HOEC)

Not all energy controls are equal. To ensure the best protection, we need a way to determine which controls are preferable when multiple options are available. The original Hierarchy of Controls (HOC) has served the safety profession for decades, ranking safety measures by their relative effectiveness as follows:

1. Elimination: Physically removing the hazard.
2. Substitution: Replacing the hazard with a less dangerous one.
3. Engineering Controls: Isolating people from the hazard using physical changes to the workplace.
4. Administrative Controls: Changing the way people work through policies, procedures, and training.
5. Personal Protective Equipment (PPE): Protecting workers with gear like gloves, masks, and helmets.

The HOC is taught widely in training, published in books, and incorporated into regulations by agencies like OSHA and NIOSH. However, it was not created through empirical research. It was created using consensus, not science, and its levels are rules of thumb, not theorems.[13]

While the HOC has become deeply embedded in the safety profession, it has limitations, especially when applied to the control of high-energy hazards. Its categories are not always distinct, and its ordering does not reflect modern safety science or the specific needs of SIF prevention. Recognizing these gaps, we developed the Hierarchy of *Energy* Control (HOEC) to provide a more effective framework tailored to controlling high-energy hazards.

Refining the Hierarchy for EBS and SIF Prevention

Here, we will take the original hierarchy and reshape it through the lens of high-energy control. The following three key changes were made to define the new HOEC.

First, we replaced "Substitution" with "Energy Reduction." It is often challenging to determine whether risk has been reduced, increased, or merely shuffled with substitution. For instance, although substituting built-up concrete formwork with panelized forms may reduce workers' time at height, it may increase the risk of crushing injuries when the forms are positioned. Energy reduction, on the other hand, can be measured objectively and reliably reduces risk, such as making changes that lower pressure in a piping system or limit the height of a fall.

Second, we removed "Administrative Controls" as a category. As we have discussed, training, rules, and procedures are essential for supporting safety programs, but they do not directly impact the human–energy interface. Since administrative controls do not meet the 3Ts criteria, we should remove them from the HOEC entirely and recognize them as outer layers of the Safety Onion. Importantly, any control that does not meet the 3Ts, administrative or otherwise, is not included in the HOEC. Some forms of control typically referred to as "administrative" may be better labeled under energy reduction or isolation if they address the energy at the source.

Third, we reclassified "PPE" as "Energy Isolation": PPE is often placed at the bottom of the HOC because it is vulnerable to human error. However, certain forms of PPE, such as arc flash suits, can function as highly effective barriers when used to isolate workers from hazardous energy. When PPE meets the criteria for a Direct Control, it should be classified as energy isolation in the HOEC.

The refined HOEC provides a framework for managing high-energy hazards (see Figure 5.2) that is based on the following theorems:

1. Energy Elimination: There is no possibility of a SIF if the energy is completely removed. *Examples: Redesigning so work at height is no longer required; de-energizing a circuit in a cabinet, verifying with a zero-voltage test, and locking the cabinet.*
2. Energy Reduction: A SIF is no longer the most likely outcome if the energy is reduced to below 1500 Joules. *Examples: Reducing pressure in a piping system during active work; a fall protection system that limits free fall.*
3. Energy Isolation: The likelihood of a SIF is negligible when the human is physically isolated from the energy. *Arc flash suit; respirator.*
4. Alternative Controls: If the energy cannot be eliminated, mitigated, or isolated, a system of alternative controls must mitigate human error to an acceptable level. Examples: *Spotter and cones; proximity detection and specialty signage.*

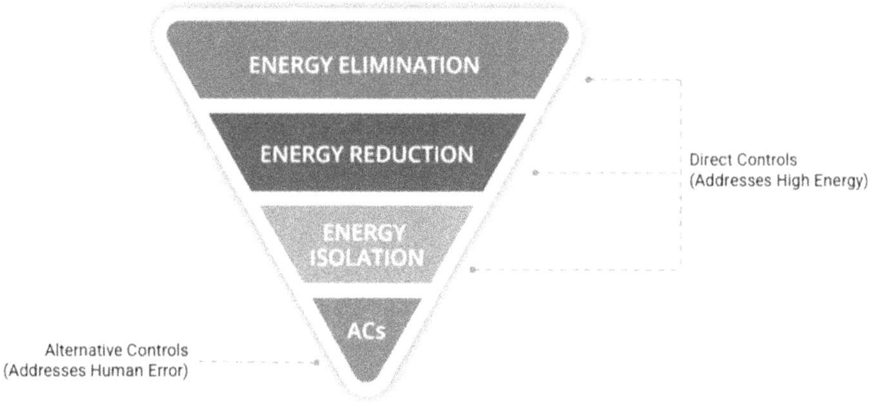

FIGURE 5.2 Hierarchy of energy control (HOEC). The HOEC prioritizes controls that eliminate, reduce, or isolate energy. These Direct Controls are most effective. When not feasible, Alternative Controls provide less reliable protection.

The HOEC provides a clear path for prioritizing energy controls: first, eliminate the energy; if elimination is not possible, reduce it; if reduction is not practical, isolate it; and if none of these are feasible, apply multiple complementary controls to address human error. The first three levels (elimination, mitigation, and isolation) constitute Direct Controls, which are always preferred, in that order. Alternative Controls, while less robust, are essential placeholders when Direct Controls are not feasible.

As with any framework, the HOEC should be a guiding tool rather than a set of strict preferences. For example, energy reduction may not be sufficient if the residual energy is still significant enough to result in recordable injuries (e.g., between 500 and 1500 Joules). In such cases, combining energy reduction with isolation may be necessary to achieve acceptable levels of risk. Similarly, organizations may prioritize isolation over reduction in some cases, depending on the specific hazard, reliability of the control measure, and availability of resources. For these reasons, the HOEC is a guideline, not a strict rulebook. As the saying goes, "All models are wrong, but some are useful." The HOEC is no exception.

The table below evaluates the same safety measures from Table 5.3, now categorizing them based on the HOEC. The safety measures are presented in their order of preference in the HOEC.

COMBINING HIGH-ENERGY CONTROL EVALUATION INTO ONE MODEL

The relationship between Direct and Alternative Controls can be confusing, with many categories, rules, and criteria. The flowchart in Figure 5.3 was created to simplify the evaluation process for high-energy hazards by consolidating the key definitions, rules, and criteria from this chapter into a single, practical tool.

The flowchart begins with a parent node: identifying the presence of a high-energy hazard to maintain focus on SIF prevention. If no high-energy hazard is present, the analysis is out of scope. However, for each high-energy hazard, controls may be evaluated against a structured series of criteria.

TABLE 5.3
Classification of Controls in the New Hierarchy of Energy Control (HOEC)

Safety Measure	Direct Control?	Type
De-energize and lock-out tag-out	Yes	Elimination
Permanent guardrail at height	Yes	Elimination
Fall protection system	Yes	Reduction
Load limiting devices on cranes	Yes	Reduction
Temporary machine guarding	Yes	Isolation
Arc flash PPE	Yes	Isolation
Spotter for an excavator	No	Alternative Control Element
Proximity detection system	No	Alternative Control Element
Cones with caution tape	No	Alternative Control Element

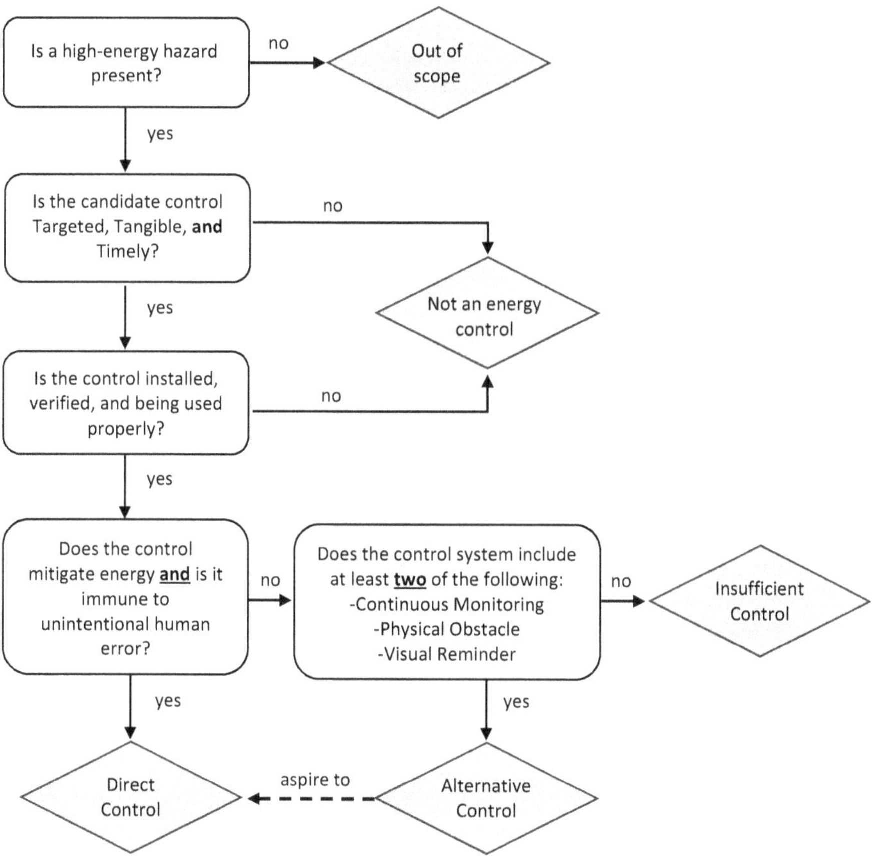

FIGURE 5.3 Model for assessing high-energy controls. This flowchart simplifies the evaluation of controls for high-energy hazards using the 3Ts criteria and definitions of Direct and Alternative Controls.

The first step is determining whether the control meets the 3Ts while ensuring it is installed, verified, and used properly. If any of these conditions are not met, the flowchart concludes that the measure is not an energy control and is, therefore, insufficient for a high-energy hazard. From there, the flowchart assesses whether the criteria for Direct Controls are satisfied, which is the desired outcome. If not, the focus transitions to evaluating the criteria for Alternative Controls. Energy controls meeting these criteria are classified as Alternative Control systems, with the goal of implementing a Direct Control in the future. If neither set of criteria is met, the control is deemed insufficient.

CONCLUSION

Research shows that high-energy hazards cause SIFs, and controls save lives. This makes the control of high-energy the north star of SIF prevention. While controls have been part of safety practices for nearly a century, the pursuit of SIF prevention has revealed gaps in how we define, assess, and categorize them.

Eliminating SIFs will require strong collaboration. To work together effectively, we need a shared approach. Standardization provides the focus and discipline needed for SIF prevention, allowing for objective assessments of energy and strict application of rules when selecting controls. Key principles such as targeting controls to specific hazards, ensuring they are observable and verifiable, and designing them to remain effective even when human error occurs are foundational to EBS.

We have begun to shift how we think about safety by emphasizing the presence of safeguards and recognizing that even the best people make mistakes. By embedding these principles into the definitions of Direct and Alternative Controls, we provide a practical way to apply HOP principles in the field, with a laser focus on SIF prevention. For those seeking to implement these principles, focusing on controlling high-energy hazards is a clear starting point.

As we pursue Direct and Alternative Controls, it is important to remember that their effectiveness hinges on human competence during design, installation, and verification. Even the most robust controls can fail if they are not installed properly or if critical steps, such as tightening a bolt, are overlooked. Alternative Controls, which rely heavily on human awareness and execution, are particularly vulnerable to error. Ultimately, the context of the work environment, including factors such as productivity pressure and distractions, shapes our ability to control energy effectively. The next chapter will delve deeper into human factors, exploring how cognitive, behavioral, and organizational elements influence energy control.

KEY TAKEAWAYS

1. Multiple studies have shown that SIFs are prevented by the Direct Control of high-energy hazards.
2. Direct Controls must be targeted, mitigate the energy, and work even if someone makes a mistake. These controls are the gold standard in SIF prevention.
3. Although having a Direct Control for every high-energy hazard is the goal, Alternative Control systems serve as a temporary solution when a Direct Control is not feasible.
4. Alternative Control systems must include at least two independent and complementary controls that reduce the likelihood of error to an acceptable level.
5. The hierarchy of energy control (HOEC) prioritizes elimination, reduction, and isolation of high-energy based on scientific theorems.

NOTES

1 Haddon Jr, W. (1995). Energy damage and the 10 countermeasure strategies. 1973. *Injury Prevention*, 1(1), 40.
2 Barnett, R. L. (2020). On the safety hierarchy and hierarchy of controls. *American Journal of Mechanical Engineering*, 8(2), 61–68.

3 de Ruijter, A., & Guldenmund, F. (2016). The bowtie method: A review. *Safety Science*, 88, 211–218.
4 Newman, A., Donohue, R., & Eva, N. (2017). Psychological safety: A systematic review of the literature. *Human Resource Management Review*, 27(3), 521–535.
5 Even the most robust energy controls depend on correct installation, verification, and use. All components must be intact, functional, and designed to withstand normal operating conditions. A competent person must verify that the control is structurally sound and free from defects, and workers must use it properly for it to be effective. Without precise implementation and oversight, even the best-designed controls may fail to perform as intended.
6 Hallowell, M. R., & Spencer, C. (2024). Safety classification & learning model: Defining & classifying potential serious injuries & fatalities. *Professional Safety*, 69(01), 18–26.
7 Conklin, T. (2019). The 5 principles of human performance: A contemporary update of the building blocks of human performance for the new view of safety. *PreAccident Media.*
8 Erkal, E. D. O., & Hallowell, M. (2023). Moving beyond TRIR: Measuring & monitoring safety performance with high-energy control assessments. *Professional Safety*, 68(05), 26–35.
9 Credit goes to my 9-year-old son, Rowan.
10 DeMott, D. L. (2014, January). Human Reliability and the Cost of Doing Business. In *Annual Maintenance and Reliability Symposium* (2014) (No. JSC-CN-31348).
11 Sameera, V., Bindra, A., & Rath, G. P. (2021). Human errors and their prevention in healthcare. *Journal of Anaesthesiology Clinical Pharmacology*, 37(3), 328–335.
12 Actual error rates and their reduction are very challenging to measure and the literature in this domain is currently very sparse.
13 Barnett, R. L. (2020). On the safety hierarchy and hierarchy of controls. *American Journal of Mechanical Engineering*, 8(2), 61–68.

6 Human Factors of Energy Control

Principle: People drive performance
Practice: Understand how context impacts energy control

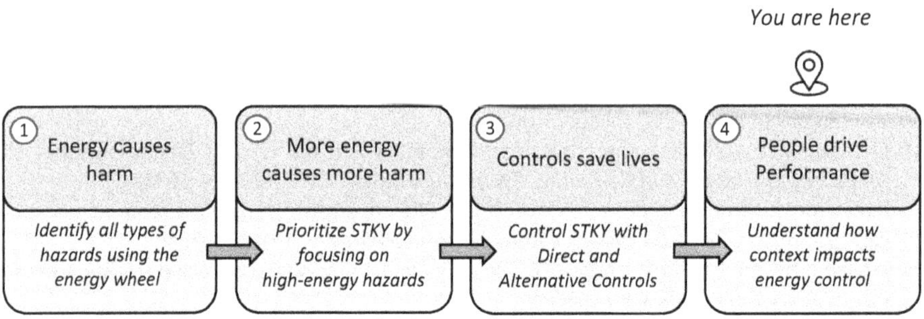

Building on extensive evidence, we have established that a SIF-focused safety program must prioritize the control of high energy. Although technically robust, all energy controls ultimately rely on humans for proper operation. Direct Controls, such as engineered barriers, are designed to function even if someone makes a mistake on the job. However, they still depend on the execution of the people who design, implement, and verify them. Any errors in these human-led functions can lead to catastrophic consequences, especially when dealing with high energy. Additionally, Alternative Controls, such as spotters and signage, are inherently vulnerable to human error, which can be intensified by contextual factors such as fatigue, distraction, and productivity pressure. As we go further down the hierarchy of energy control (HOEC), controls may be easier and less expensive to implement, but they depend more and more on human perfection. Therefore, preventing SIFs requires a deep understanding of human error and the factors that drive it.

Before diving into the mechanisms of human error, it's important to address a key misconception about error itself. For years, people have been blamed, shamed, and retrained because they made an error. When we punish people for human error, we are punishing people for being human. It's unfair, and it doesn't improve the safety of future work. Instead, it creates an environment of fear, decimates trust, and stifles learning.[1]

Thanks to the contributions of Dekker, Conklin, and others, the safety profession is beginning to correct decades of misguided behavior, exchanging blame for learning. It is now widely recognized that human error is almost never the root cause of an incident because countless contextual factors precede and explain human actions.

 DOI: 10.1201/9781003641483-7

However, while this progress is commendable, it has sometimes led to an overcorrection. Safety professionals sometimes swing the pendulum too far, focusing exclusively on systems and overlooking the role of human error entirely.[2] To effectively address SIFs, we must move beyond a binary view and examine the dynamic interplay between people and systems. Error is not a dirty word, and ignoring its role in SIFs means turning a blind eye to half of the equation.

This chapter builds on three related principles of Human and Organizational Performance (HOP): error is normal and to be expected; context drives behavior; and how we respond to failure matters.[3] By understanding and improving the context in which people work, we can minimize the factors that contribute to errors, significantly enhance the reliability of energy controls and, ultimately, prevent SIFs.

As we proceed, we will work backwards from the fundamentals of human error to the identifiable and actionable precursors that precede them. The architecture of this discussion is illustrated in Figure 6.1. Precursors and errors are intrinsically connected. While errors explain the cognitive reasons why controls may be improperly implemented, precursors provide the contextual reasons behind those errors. By understanding this connection, we can better explain how precursors translate to SIFs.

Before examining actionable precursors, we will delve into the mechanics of human error, answering questions like: *What is an error? What are the various types of errors?* and *Why do errors occur?* This foundational knowledge will be used to show how error is an inevitable part of everyday life. It will also serve as a framework to explain why specific work factors like work pressure influence behavior and can be used as identifiable precursors of uncontrolled high-energy hazards.

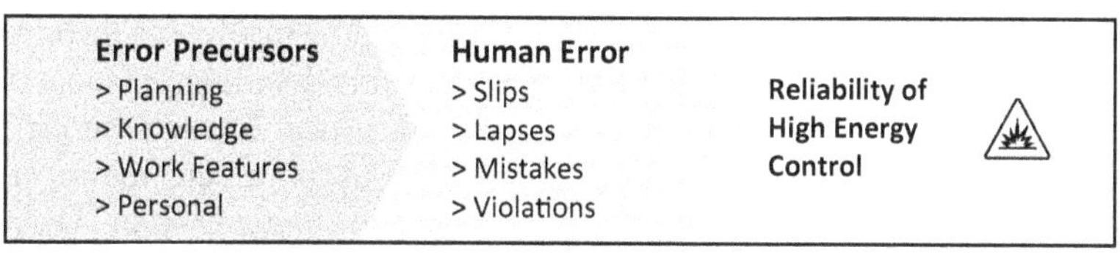

FIGURE 6.1 Relationships between error precursors, human error, and the reliability of energy control. This model illustrates how error precursors increase the likelihood of human error, which in turn undermines the reliability of energy controls.

HUMAN ERROR: UNDERSTANDING THE FUNDAMENTAL DRIVERS OF HUMAN PERFORMANCE

To create a work environment conducive to effective energy control, we must first understand how and why errors occur. James Reason offers a classic model for understanding human error. Rooted in cognitive psychology, his model – visually adapted in Figure 6.2 – is both balanced and practical, emphasizing the interplay between individuals and their environment. Instead of drawing abstract conclusions, his model points us to specific actions that improve our chances of success.

Although many scholars have explored and commented on human error, James Reason's work is widely regarded to be foundational due to its strong grounding in cognitive psychology. In the first part of this chapter, I have tried to summarize the key elements of his model, which guide us in identifying and addressing error precursors.

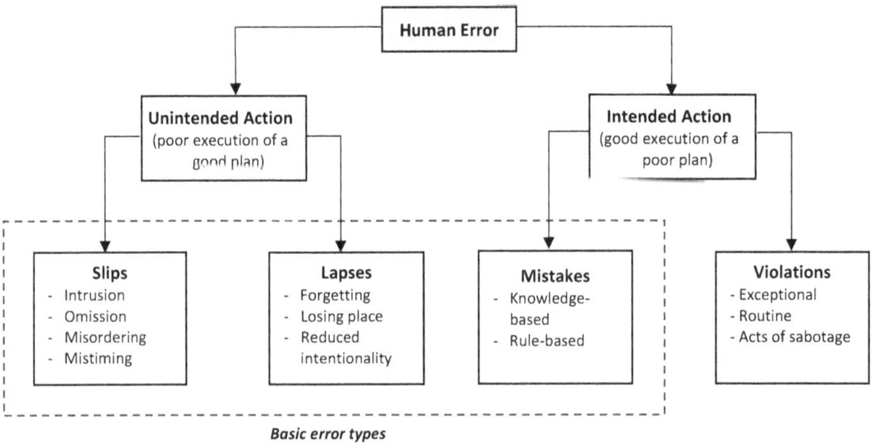

FIGURE 6.2 James Reason's human error classification model. Adapted from Reason's work, this model categorizes errors into slips, lapses, mistakes, and violations, providing a foundation for understanding how different types of human error affect control reliability.

Reason, J. (1990). *Human Error*. Cambridge university press. Figure 7.7, page 207.

What Is an Error?

The first step in using Reason's model is to define and identify errors. He defines an error as "a deviation from intended actions or decisions, which can result in unintended outcomes." Practically, I think of error as something that we would have done differently with the benefit of hindsight. Note that this definition is about action, not outcome. Decoupling actions from outcomes is critical because we can make many errors, sometimes without knowing it, before a negative outcome occurs. For example, failing to check your vehicle's blind spot before merging would be an error (action), but getting in a car crash would not (outcome). We can't find and fix error-prone work situations if we can't identify where and when errors are occurring.

ERROR EXAMPLE – CASE OF THE SOAPY WATER

The night before a team bike ride, I decided to clean my water bottles. I put dish soap and water in the bottles and left them to sit overnight to clean. When I woke in the morning, I realized I had not set my alarm and rushed to get my things together before my teammates arrived. I grabbed both bottles without emptying and rinsing them. Halfway through the ride, I realized that I had only soapy water on a very hot day.

What would I have done differently in hindsight? There are two actions in need of correction: (1) set my alarm so I am not rushed and (2) empty and rinse the bottles. Both errors contributed to the outcome (having no fresh water), but they occurred for different reasons.

UNINTENTIONAL ACTIONS: EXECUTION ERRORS

It is very important to distinguish between good plans that are executed poorly and bad plans that are executed well because the drivers are very different. A common type of error occurs when we have a good plan in place, but we execute it poorly because of an unintended action or inaction.[4] That is, we fail to execute a series of actions that must be made in a particular time and order. A great example of such a situation is a chef preparing a complex meal with a recipe that requires a series of carefully timed steps such as preheating the oven, sautéing ingredients, and plating dishes where a single missed step or mistimed action can ruin the entire meal.

Unintended actions often occur when expert workers perform typical tasks in routine environments. They occur when we go on "autopilot," particularly when we are distracted or fatigued. These human factors compromise the cognition of even the most highly skilled and trained professionals.

There are generally two categories of unintentional actions: slips and lapses. Slips typically occur when attention naturally falters, leading to an incorrect or unintended action, while lapses are usually the result of memory failures, such as forgetting to perform a necessary step.

SLIPS

There are four general examples of slips, described below.

1. *Intrusion*: Following custom of practice in a new situation. This often occurs when there is a change to an otherwise standard process.
   a. Everyday Example: Returning to a previous address after moving across town.
   b. High-Energy Control Example: The company purchases new equipment that requires a different de-energization procedure. An employee who has been trained on the new process follows the old process instead.
2. *Omission*: Leaving out a required step of a process.
   a. Example: Cooking Pad Thai and not including the egg.
   b. High-Energy Control Example: A roofer gathers all his fall protection equipment, correctly wears his harness, but does not secure the harness to the anchor point.
3. *Misordering*: Executing required steps in an incorrect sequence.
   a. Everyday Example: Starting the car, putting the car into reverse, and hitting the accelerator before releasing the emergency brake.
   b. High-Energy Control Example: Crane operator begins to swing a load before establishing vertical clearance.
4. *Mistiming*: Executing a required step at the wrong time.
   a. Everyday Example: Missing an exit after signaling and changing lanes too late.
   b. High-Energy Control Example: Removing temporary structural supports before concrete has adequately cured.

Because slips are related to work execution rather than planning errors, they do not result from poor training or a lack of expertise. Therefore, retraining workers is an inappropriate response to slips. If someone can do the work correctly when asked, then it isn't a training issue. Instead, these errors typically arise from context (i.e., situational factors in the environment in which we operate). Proper countermeasures include ensuring employees are well-rested, not rushed, free from distractions and interruptions, and having a second set of eyes on critical steps.

Lapses

Lapses are memory failures that occur during a task when an intended action is forgotten. Unlike slips, which are execution errors, lapses involve failing to recall or remember specific actions or steps. These errors occur even when people know what to do and mean to do it. They're especially dangerous for high-energy work that requires quick decisions.

Memory lapses are so pervasive that "forgetting" is acknowledged as a normal part of everyday life. Everyone forgets things. Just today, my wife forgot the garage code, my son forgot his socks, and I blanked on an old teammate's name. Usually, memory failures cause minor inconveniences. However, they can sometimes lead to serious consequences, like when I forgot to secure my bike to the roof rack, and it shattered my back window while driving on the highway.

Memory failures can occur at any time and without warning, making them difficult to prevent. Effective solutions include using checklists, taking regular breaks, being well rested, and minimizing distractions and interruptions.

INTENTIONAL ACTIONS: PLANNING ERRORS

Another important family of errors are intentional actions related to failures in planning. At first, the idea of intentional errors can seem counterintuitive. Who would *intend* to make an error? However, as we begin to understand errors as actions and not outcomes, it becomes clear how we often actively choose the wrong plan and act accordingly.

Sometimes we take an action that we didn't realize was incorrect, while other times, we knowingly choose a poor plan and proceed anyway. Take the soapy water example. Why didn't I set my alarm? I knew better, and it was a conscious decision. I thought, "I'll be fine; I always wake up with plenty of time." The action was intentional, and the next morning, I regretted it. I had violated my own unwritten rule. We make these sorts of intentional errors more often than we might like to admit. Although contrary to some safety beliefs, we must understand that sometimes people do choose the wrong plan, even if it's not their fault.

Intentional errors fall into two primary categories: mistakes and violations. These planning errors highlight the complexity of human decision-making and the many factors that can result in undesirable decisions.

Mistakes

Up until this point, we have casually used the word mistake as a synonym for error. However, in the field of cognitive psychology, the word mistake has a specific meaning. Unlike unintentional errors such as forgetting, mistakes are intentional actions. Although rarely malicious, mistakes involve failures in the planning process, typically resulting from insufficient knowledge or inadequate procedures. Mistakes are particularly common when novice workers are still learning correct procedures or when experienced workers are put into unfamiliar situations.

Mistakes generally fit into one of the following three categories:

1. *Knowledge-Based Mistakes*: Following an incorrect plan due to inadequate knowledge or understanding of a situation or process. Typically, we follow an error-driven process to learn the correct plan where we make a series of mistakes and corrections to hopefully achieve a satisfactory outcome.
   a. Everyday Example: Trying to ski for the first time without lessons by pointing skis straight down the mountain. If you don't know how to "pizza," you're gonna have a bad time.
   b. High-Energy Control Example: Mixing two highly reactive chemicals due to insufficient training or experience.
2. *Misapplication of a Good Rule*: Correctly following a generally valid rule or procedure but applying it in an inappropriate context or situation. Not every rule or procedure is ideal all the time, so it is important to understand the exceptions.
   a. Everyday Example: Using the "righty-tighty, lefty-loosey" rule to unscrew the left pedal of a bicycle. Since it is reverse-threaded, this rule will only tighten it further instead of loosening it.
   b. High-Energy Control Example: Following a rule to use a self-retracting lifeline (SRL) when there is insufficient clearance below for the SRL to engage properly.
3. *Application of a Bad Rule*: Implementing a rule or procedure that is inherently flawed or inappropriate.
   a. Everyday Example: Using hot water to wash all types of clothing because you believe it cleans better, even though it can damage delicate clothes.
   b. High-Energy Control Example: Using the same standard lock for all lock-out tag-out (LOTO) procedures, even when different equipment needs specific locks for complete isolation.

Mistakes can be predictable. People make knowledge-based mistakes in unfamiliar environments and make rule-based mistakes when procedures are incorrect or incomplete. To prevent SIFs, new employees must be well trained before they take on tasks that involve high-energy hazards. Furthermore, when work configurations change, we must verify that even our most experienced employees understand the hazards and how to apply the necessary controls.

Violations

The final category of error is perhaps the most controversial. Violations are intentional deviations from established procedures, rules, or standards. Unlike mistakes, which typically stem from a lack of knowledge or a poor procedure, violations occur when individuals knowingly choose to disregard a good procedure. In other words, we know better, but we do it anyway.

Does this mean that people who commit violations are bad actors? Generally, no. Violations are most often driven by factors such as schedule pressure, overconfidence in one's abilities, or a belief that the rules are unnecessary or overly restrictive.[5] Violations can be categorized as routine violations, which have become normalized in an organization, and exceptional violations, which only occur in extraordinary circumstances. Violations can include subtle actions like purposely skipping a step in a procedure or egregious like deliberately defeating a control.

When a violation occurs, it is critical to understand and address the reasons behind it rather than simply punishing the individual. In the case of routine violations, a bigger problem may indicate issues with the company's values, norms, and management systems. While the idea of discipline is less popular in some safety philosophies, there are still cases where serious, intentional violations may warrant strong consequences. Regardless of the chosen response, violations should be viewed as an opportunity for learning.

ERROR CLASSIFICATION AND ITS ROLE IN PREVENTION

Classifying errors helps us identify why errors occur and what can be done to prevent them. By examining specific cases, we can better understand how different types of errors arise and design targeted countermeasures.

The soapy water example illustrates two distinct classifications of error arising from one case, each with its own contributing factors and potential solutions. The first error, not setting the alarm, was a routine violation. Overconfidence in my ability to wake up without an alarm led to a conscious decision not to set one. The second error, not rinsing the bottles, was an intrusion caused by distraction and habitual actions in a rushed situation. Together, these classifications demonstrate how both intentional and unintentional errors can contribute to the same negative outcome.

ERROR 1: NOT SETTING THE ALARM

- *Did you intentionally not set your alarm?* Yes, I consciously decided not to set my alarm thinking that I would wake up on time.
- *Is there a "rule" to set your alarm?* Yes, I generally have my own "rule" to set an alarm if I have something important to do in the morning.
- *Do you not set an alarm often?* Yes, and I should probably stop doing that.

Classification: Routine violation.

Countermeasure: Improve my habits with setting alarms or, better yet, buy an alarm clock that doesn't require me to manually reset my alarm each morning.

ERROR 2: NOT RINSING THE BOTTLES

- *Did you not rinse the bottles on purpose?* No, I would have rinsed these if I had been thinking straight.
- *Was there a distraction present?* Yes, I was super rushed and frazzled because I was late.
- *Did you follow a typical process in a new situation?* Yes, I usually fill my bottles before I ride but I rarely put soap in them and let them sit overnight. I grabbed the bottles out of habit.

Classification: Intrusion.

Countermeasure: Label the bottles as a reminder that there is soap, leave the caps off the bottles while they disinfect so the bubbles are visible, or finish cleaning the bottles the night before.

PRECURSORS OF ERROR: HOW CONTEXT DRIVES BEHAVIOR

While classification is invaluable for understanding why errors happen, it is only part of the story. To truly prevent errors, we must examine the conditions that create error-prone situations. As James Reason eloquently stated, "Rather than being the main instigators of an accident, operators tend to be the inheritors of system defects created by poor design, incorrect installation, faulty maintenance, and bad management decisions. Their part is usually that of adding the final garnish to a lethal brew whose ingredients have already been long in the cooking.[6]" Similarly, as Conklin elegantly states, "context drives behavior." Therefore, to prevent SIFs, it is critical to understand what "context" truly entails, how it shapes human behavior, and what can be done to diagnose and improve it in practice.

In the next section, we will explore the specific factors that contribute to errors related to the control of high-energy. After unpacking them and reviewing how they can be identified through field safety engagements, we will make specific connections to the error types showing, for example, that the quality of planning relates directly to planning errors like mistakes and violations, and how the features of work like productivity pressure and fatigue relate to unintentional slips and lapses. Understanding this complex interplay will help us better address human factors that are essential to reliable energy control.

The Precursor Studies

The factors that contribute to errors are well known and comprehensively documented.[7] Situational influences such as long work hours, time pressure, and distractions define the "context" that shapes human behavior. When these influences are identifiable, predictive, and actionable, we refer to them as error precursors.

Over the course of a decade, we studied these factors extensively using a technique called precursor analysis.

Precursor analysis, a method for identifying and correcting the factors that lead to catastrophic outcomes, has long been a cornerstone of safety in high-risk industries like nuclear energy, aviation, and chemical manufacturing. Its origins trace back to the Three Mile Island nuclear plant meltdown in 1979, which spurred the development of rigorous programs to identify and mitigate error precursors. Similar initiatives, such as the Aviation Safety Action Program (ASAP) and NASA's Accident Precursor Analysis (APA), followed suit, each tailored to its respective industry's needs. Despite these variations, the core objective remains the same: to pinpoint a small number of predictive conditions that can reliably forecast and prevent catastrophic events.

Building on this precedent, my esteemed PhD advisor, John Gambatese, and I sought to adapt these methods to SIF prevention in construction, operations, and maintenance work.[8] By identifying the key precursors of SIF-related error, this work helped make the principle that "context drives behavior" more tangible and actionable.

Our study had three main steps.

1. **Identifying potential precursors**
   Our first step was to compile a comprehensive list of potential precursors through a literature review, root cause analyses, and expert brainstorming. We focused narrowly on factors that were identifiable before work began, predictive of SIFs, and actionable. This yielded over 50 potential precursors, a valuable but overwhelming list.
2. **Shifting from a retrospective to a proactive paradigm**
   Not all causal factors qualify as precursors. For a factor to be considered a true precursor, it must meet three critical criteria: proactive, predictive, and identifiable before the incident occurs. To put these criteria into practice, we developed an engagement protocol that shifted the focus from hindsight to foresight. Instead of relying on post-incident analysis, we collected contextual information in real time, engaging workers with open-ended questions designed to capture meaningful insights. For example, instead of asking, "Do you know the safe work procedure?" we asked, "Can you tell me about your work today and how you plan to complete it safely?" This approach allowed us to uncover predictive conditions before errors or incidents occurred, firmly embedding precursors within a proactive safety paradigm. The suite of questions will be shared later in this chapter as we discuss implementation.
3. **Collecting and analyzing empirical data**
   Using the engagement protocol developed in Step 2, we focused on collecting real-time data to identify and validate error precursors. Across three studies, we gathered over 150 cases, evenly split between "success" (high-energy work completed with effective controls) and "failure" (high-energy work where controls were absent, leading to energy release). By comparing

success and failure cases, we identified contextual factors that reliably predicted the likelihood of a SIF. These factors represent the key precursors that signal increased risk, allowing for proactive intervention.

REVIEW OF KEY PRECURSORS

More than 50 factors were studied and, through cross-industry experiments and rigorous statistical analysis, narrowed to 11 precursors that are identifiable, actionable, predictive, and apply across industry types and sectors.[9,10] These precursors explained actions such as omitting a required control, selecting the wrong control, or failing to use a control correctly. By narrowing the focus from over 50 potential factors to 11, we not only clarified the path forward but also made precursor analysis feasible for practical applications. What once required hours of engagement can now be completed in about 10 minutes, enabling us to quickly assess and address the conditions that contribute most to energy control. These findings have since been validated across multiple sectors, including construction, electric utility work, and elevator installation, demonstrating their universal relevance and applicability.

The 11 key precursors are listed below. Surely, many readers will review this list and immediately think of several factors that appear to be missing. Keep in mind that our research team brainstormed a list of over 50 factors, but the research showed that 80% of these are not predictive. It doesn't mean that these other factors aren't important and shouldn't be assessed, but these 11 pass the high bar of being statistically significant when the others did not. Put simply, you can assess whatever factors you wish, but don't miss these 11!

When assessing these error precursors, the engagement protocol developed during the research provides a useful starting point. For each precursor, we identified example questions that can guide discussions and uncover meaningful insights. These sample questions are intended as flexible guidance rather than a rigid interview protocol. Each of the 11 precursors will be described in detail below, along with suggested engagement questions that can help uncover the contextual factors at play. In Chapter 9, we will further explore how these precursors can be assessed through fluid discussions during typical Safety Walks.

1. **Inadequate pre-job safety brief**
   A pre-job safety brief is a safety-focused discussion held by a crew immediately prior to work. Although most mature companies hold these meetings regularly, they vary greatly in their quality. When a poor meeting occurs, hazards can be overlooked, controls can be inadequate or missing, and there can be gaps in understanding of job steps and roles. These gaps severely compromise the control of high-energy hazards because they cause both mistakes and violations. Teams are especially vulnerable when there are novice workers or unfamiliar tasks.

   It is easy to identify when a pre-job safety brief is inadequate simply by watching the meeting. In Chapter 8, we will give a full treatment of "what good looks like" for pre-job safety briefs and how to assess quality through observation. However, if you don't have the opportunity to view the meeting, a few questions to ask in an engagement are provided below.

- What hazards might you face with your work today (use the energy wheel)?
- What high-energy hazards might you face with your work today?
- What direct or alternative controls do you have in place for serious hazards?

2. **Inadequate plan to address change**

In addition to planning for work as it is imagined, we must plan for inevitable change. As the research shows, about 30% of all hazards are not identifiable before work begins and over half of all SIFs involve last-minute changes that lead to loss of energy control.[11] Although it is not possible to anticipate all potential changes as we discussed earlier, crews must have a plan for what to do when something does change. Such a plan discourages improvisation and encourages the use of stop work authority. Deficiencies in this planning are apparent when there is little to no discussion about change and stop work, or when there is a general overconfidence that work will proceed as imagined. Again, this factor is best assessed by observing the pre-job safety brief but may also be assessed after by asking some of the following questions.

- How might the work or work environment change?
- How will you manage these changes if they occur?
- When was the last time your crew stopped work for a safety concern?

3. **Insufficient knowledge of safe work procedures**

Safe work procedures can be a roadmap to successful outcomes for high-energy tasks, especially when the work is complex and there is a particular set of steps that must be strictly followed. Most companies maintain documents that describe safety protocols for these tasks and provide specialty training to their employees. To be effective, even the best protocols must be carefully read and understood. Understandably, this does not happen when employees are inundated with volumes of paperwork. We often greatly overestimate how much people can (or are willing to) read and remember.

When assessing this factor, it is important to be subtle. As mentioned, simply asking for confirmation that the employee is aware of the safe work procedure does not indicate whether they understand it. If we asked a crew "*do you know the safety work procedure?*" who would ever say "*no*"?

Therefore, we recommend the simpler and more open questions provided below. To be effective, the leader who is performing the engagement must know the safe work procedure well enough to judge the employee's understanding.

- What are you working on today?
- Who is doing each task?
- What is the safe work procedure for the task today?

4. **Departure from routine**
Minor but important changes in routine are triggers for human error. For example, a seemingly trivial change in location of a valve or slight reordering steps in a lock-out tag-out process can easily lead to lapses in judgment, especially intrusions where people follow custom of practice in a new situation. In hindsight, these actions may seem like complacency or ignorance, but they are actually quite predictable and understandable from a human perspective. Therefore, when a routine has changed related to the control of a high-energy hazard, we should be hyper vigilant about ensuring the new procedure is followed.

- What about the work might not be completed as planned?
- What is different about your work today (even little stuff)?
- What part of this job do you expect to troubleshoot?

5. **Unfamiliar with the task**
When a task is familiar and a safe work process is repeated many times, our brain creates mental models and neural pathways that support the routine. This is important with high-risk work because we are more likely to defer to the safe routine when faced with distractions and fatigue. If we have not developed good routines, we are much more susceptible to knowledge-based mistakes, slips, and rule-based violations. Familiarity with a task is almost impossible to train in a classroom; instead, it requires repeated actions in actual work situations. Familiarity with a task can be assessed by learning more about an employee's history with the high-energy task and associated controls.

- How long have you been doing this task?
- Since the first time, how has the task been changed?
- When was the last time you performed this task this way?

6. **Productivity pressure**

Work pressure is normal and ever present. After all, the faster the company delivers a quality product, the more money the company will make. However, there are times when work pressure is unusually high. For example, there might be unusual pressure if a crew is playing catch-up after a delay; stress from a client who is displeased with the pace of work; or a task involves a very expensive piece of equipment. When asking the following questions, attention should be placed on pressures that are *unusual or extreme*. You will find that employees are often very willing to explain how the actions of the company, clients, or public cause them undue stress, and they tend to be very thankful that a leader is listening. For this assessment, workers' *perceived* productivity pressure is reality because it influences behavior. Management perception of actual pressure is far less relevant.

- Have you had any work disruptions lately (like delays or events)?
- Once the job is complete, what's next?
- What would happen if the job wasn't completed today?

7. **Long work hours**

I am often asked "how many hours is too many?" The research shows that it depends on what is normal for the crew. For example, if a crew is accustomed to 12-hour workdays, then a 12-hour day may not significantly impact performance. However, if the standard is an 8-hour workday, suddenly requiring a 12-hour shift can be disruptive, both physically and mentally. From an error perspective, long hours reduce alertness, which increases the likelihood of slips, lapses, and mistakes. Additionally, when long hours are imposed rather than voluntary, they can lead to frustration, causing exceptional violations as workers become desperate to finish the job and go home. The good news is that employees are typically open to sharing how extended hours affect them, especially when asked in a supportive and non-judgmental manner.

- What are your normal work hours?
- What hours are you working today?
- How often are you asked to work much longer than you would prefer?

8. **Distractions**

Distractions are any factors that divert attention away from the task at hand and, when significant, can lead to unintentional errors. They can be physical, such as a co-worker interrupting focus, or emotional, such as the stress of a negative relationship with a supervisor. Distractions often extend beyond the workplace. For instance, stress from an argument with a spouse or concern for a sick child can be even more distracting than work-related issues. Interestingly, distractions can also stem from positive life events, such as the excitement of a new baby or the pride of buying a dream car. Before asking employees about their distractions, it is essential to build positive rapport and approach the conversation with tact and respect, even if you have a strong relationship with them. Companies with high levels of trust tend to receive richer and more honest answers, making it easier to address these factors effectively.

- What is on your mind today?
- What about the work today might be distracting?
- What other things might be distracting (from outside work)?

9. **Fatigue**

Fatigue is a complex phenomenon that involves both mental and physical exhaustion. Fatigue is well known to cause cognitive and physical impairment, increasing the frequency of execution errors such as slips and lapses. For example, people are most likely to forget a step or lose track of a sequence at the end of the workday.

Fatigue is notoriously difficult to measure because it is caused by a host of personal and work-related factors. For example, the following factors highly impact fatigue: quality of sleep, noise, temperature, age, and whether work hours were imposed or selected. Like investigating personal distraction, we must carefully craft our questions about fatigue.

- What might make you fatigued today from your work?
- What else might make you fatigued today?
- How have you been sleeping?

10. **Risk normalization**

Risk normalization is the gradual acceptance of unsafe practices as normal. As humans, repeated exposure to hazards without negative outcomes dulls our fear. Think of your first time climbing a tall ladder; it may have been both exciting and scary. Over time, after using ladders repeatedly, that fear fades. This drift happens with all hazard types and is incredibly challenging to counteract.

Seasoned employees with years of experience working around specific hazards are particularly vulnerable to risk normalization. For example, linemen may no longer feel fear toward high voltage, and elevator mechanics may forget the danger of a one-ton counterweight moving just a foot away.

As we grow comfortable around high-energy hazards, the fear that helps keep us safe diminishes, and our control over that energy can drift as well. Normalization often leads to overconfidence, lost focus, and shortcuts near high-energy sources. While we may not be able to change someone's tolerance for risk, we can remind them of expected work procedures and required controls.

- How long have you worked around these hazards?
- What makes you nervous about your work?
- What critical step in the work procedure could go wrong?

11. **Poor attitudes toward safety**

Unlike other precursors that require specific questions or observations to assess, attitudes can be observed throughout the entire engagement. For instance, some individuals may be despondent, display a lack of awareness about hazards and controls, or dismiss the company's policies and procedures. When prevalent across a group, poor safety attitudes may indicate a larger systemic issue that needs to be addressed. However, if negative attitudes are expressed by one individual rather than the group, the issue may warrant a targeted correction.

It is crucial to distinguish displeasure with the company or constructive feedback from genuinely poor safety attitudes. Suggestions for improvement, even if critical, are valuable and should be welcomed. An open and assertive workforce willing to identify areas for improvement is far better than a quiet and compliant one. If employees are unwilling to share their insights or concerns, that reluctance may signal a deeper problem.

- How do you contribute to the safety of yourself and your co-workers?
- How could safety be improved?
- What lessons learned have you shared with co-workers and the company?

CONNECTION BETWEEN PRECURSORS AND ERROR

Precursors and errors are intrinsically connected. Perhaps most obvious, the quality of pre-job safety briefs directly relates to planning errors. For example, if a crew has a poor pre-job safety briefing, some newer employees may not understand their role or may make assumptions that lead to poor knowledge-based mistakes. Similarly, precursors related to the features of the work like productivity pressure and long hours may lead directly to unintentional errors like slips and lapses.

In Table 6.1, the 11 precursors are grouped into four families to highlight their logical relationships. By organizing the precursors in this way, the table provides a practical framework for understanding how contextual factors may influence different types of errors. Although errors can occur without any precursor, they are far more likely when the associated precursors are present. This table serves as a reference tool, helping to diagnose and address some conditions that contribute to SIFs. Arguments could be made that any of the precursors could impact any of the error types. However, for simplicity, I have included those that we have observed most often.

TABLE 6.1
Examples of How Precursors and Error Can Be Logically Connected

		Execution Errors		Planning Errors	
Family	**Precursor**	**Slips**	**Lapses**	**Mistakes**	**Violations**
Quality of Planning	Inadequate pre-job safety brief			X	X
	Inadequate plan to address change			X	X
Knowledge of Work	Insufficient knowledge of safe work Procedures			X	
	Departure from routine			X	
	Unfamiliar with the task			X	
Features of Work	Productivity pressure	X	X		
	Long work hours	X	X		
	Distractions[a]	X	X		
	Fatigue[a]	X	X		X
Individual Characteristics	Risk normalization				X
	Poor attitudes toward safety				X

[a] Distractions and fatigue can be related to one's personal life and work life.

CONCLUSION

This final chapter in the theory of EBS has explored how people drive performance and how human error directly impacts the control of energy. Errors, such as omitting a required control, applying the wrong one, or failing to implement it correctly, are critical because they explain how deficiencies in controlling energy can lead to SIFs. While EBS focuses on preventing harm through effective controls, this chapter has shown that focusing on the controls themselves is not enough. Ignoring the role of human error in energy control would leave a critical gap in the EBS framework. Put simply, we must understand error, and act on precursors.

Error precursors provide essential insight into why errors occur and, more importantly, how they can be prevented. These precursors highlight the contextual factors, such as fatigue, distractions, or risk normalization, that increase the likelihood that controls will be improperly implemented. By addressing these underlying conditions, we can deploy proactive strategies to mitigate risk and improve the reliability of controls, even in challenging situations.

We have learned that energy causes harm, more energy causes more harm, controls save lives, and people drive performance. As we move forward, Part II of this book will focus on the implementation and integration of these EBS principles in practice.

KEY TAKEAWAYS

1. Direct and Alternative Controls rely heavily on the reliability of the people who implement, verify, and use them.
2. Error is a normal and expected part of the human condition. James Reason's human error model provides a balanced and practical method to explain how errors occur.
3. Errors are shaped by the surrounding context, which are defined by the presence of key precursors.
4. Research identified 11 precursors that are both predictive and actionable, enabling safety professionals to identify the conditions that lead to critical errors before they occur.
5. Managing human error and precursors together strengthens energy control and makes systems more reliable.

NOTES

1 Dekker, S. (2017). *The Field Guide to Human Error Investigations*. Routledge.
2 Sherratt, F., Thallapureddy, S., Bhandari, S., Hansen, H., Harch, D., & Hallowell, M. R. (2023). The unintended consequences of no blame ideology for incident investigation in the US construction industry. *Safety Science*, 166, 106247.
3 Conklin, T. (2019). The 5 principles of human performance: A contemporary update of the building blocks of human performance for the new view of safety. *PreAccident Media*.
4 James, K. L., Barlow, D., McArtney, R., Hiom, S., Roberts, D., & Whittlesea, C. (2009). Incidence, type and causes of dispensing errors: A review of the literature. *International Journal of Pharmacy Practice*, 17(1), 9–30.
5 de Mattos, L. A., Rocha, R., & de Castro Moura Duarte, F. J. (2024). Human error and violation of rules in industrial safety: A systematic literature review. *Work*, 79(3), 1237–1253.
6 Reason, J. (1990). *Human Error*. Cambridge University Press., page 173.

7 Standard, D. O. E. (2009). *Human Performance Improvement Handbook Volume 1: Concepts And Principles*. US Department of Energy AREA HFAC Washington, DC, 20585.
8 Alexander, D., Hallowell, M., & Gambatese, J. (2017). Precursors of construction fatalities. I: Iterative experiment to test the predictive validity of human judgment. *Journal of Construction Engineering and Management*, 143(7), 04017023.
9 Alexander, D., Hallowell, M., & Gambatese, J. (2017). Precursors of construction fatalities. II: Predictive modeling and empirical validation. *Journal of Construction Engineering and Management*, 143(7), 04017024.
10 Hallowell, M. (2019). *Serious Injury and Fatality (SIF) Precursor Customization Project: Implementation Guide*. Edison Electric Institute. Retrieved from https://www.eei.org/-/media/Project/EEI/Documents/Issues-and-Policy/Power-to-Prevent-SIF/sifPrecursorGuide.pdf.
11 Lee, W. H., Gambatese, J., Hallowell, M. R., & Nnaji, C. Impacts of Last-Minute Changes on Construction Safety. In *Construction Research Congress* 2024 (pp. 497–507).

Part 2

EBS Integration

In Part I of this book, we explored EBS principles, drawing on evidence from two decades of research. We learned that energy causes harm, more energy causes more harm, controls save lives, and people ultimately drive performance. In Part II, we transition from theory to practice, showing how EBS principles provide a focused and practical framework for SIF prevention.

EBS aligns practical action with strategic priorities. It sharpens the focus on SIF within existing safety activities, such as pre-job planning and Safety Walks, by emphasizing high-energy hazards and their controls. At the same time, EBS strengthens the Safety Management System (SMS)[1] by ensuring that organizational goals, activities, and metrics align with the primary objective of controlling high energy.

Modern paradigms like HOP have transformed how we approach safety. We have shifted from viewing safety as the absence of injuries to the presence of safeguards, from blaming employees to understanding their work context, and from short-term reactions to sustained learning and improvement. Despite this progress, gaps often remain between what companies say, what they do, and what they measure. EBS bridges these gaps by operationalizing this thinking. For example, prioritizing high-energy controls in design, planning, and work execution reflects the principle that safety is the presence of safeguards; ensuring Direct Controls account for human fallibility institutionalizes the understanding that error is normal; and assessing SIF precursors during Safety Walks will illuminate the context that drives behavior. By integrating EBS principles across the SMS and into core safety activities, we also weave HOP principles into the fabric of the organization.

Although EBS may seem like a complete overhaul of the safety program, it is not. Instead, it provides a practical way to add a SIF focus to activities most companies

DOI: 10.1201/9781003641483-8

already perform. For many organizations, EBS is simply a "quarter turn" on an already strong safety system. This distinction is important because companies have invested heavily in their safety programs and made remarkable progress in injury prevention. That progress should not be abandoned. Instead, EBS builds on these systems, using them as a scaffold for SIF prevention. EBS is not another task or a "flavor of the month"; it is a way to enhance what we already do.

Part II will explore how EBS principles and new safety thinking can be embedded throughout the safety program, diving deeply into five key areas. Illustrated in Figure P2.1, these areas roughly correspond to the plan, do, check, act (PDCA) cycle.[2] By integrating EBS into each phase, organizations can address high-energy hazards systematically, ensuring continuous improvement and building a more strategic approach to SIF prevention. Figure P2.1 will serve as our roadmap for Part II of this book, with the subsequent chapters covering each step in the cycle.

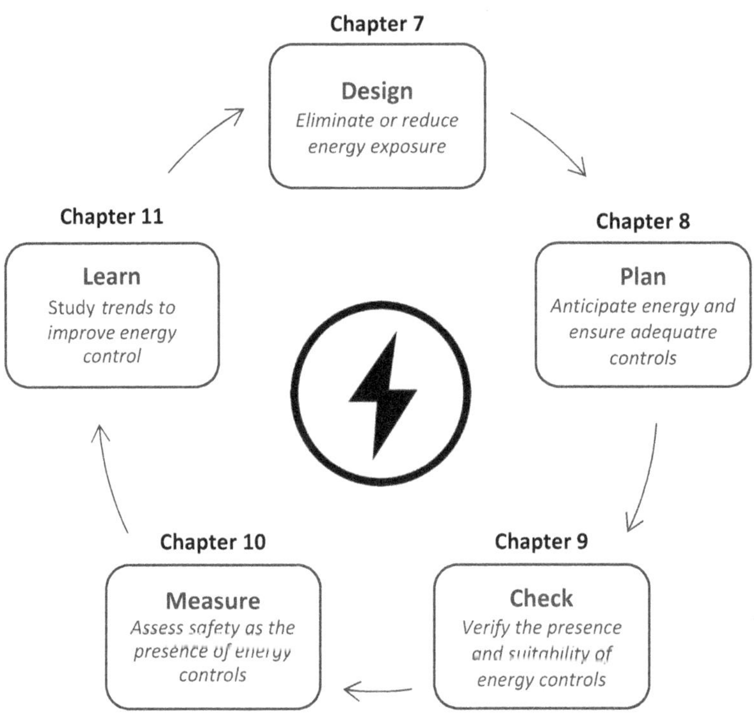

FIGURE P2.1 Approach to aligning the safety management system with EBS principles. Our roadmap for Chapters 7–11.

NOTES

1 For more information about SMS, see International Organization for Standardization. (2018). ISO 45001:2018: Occupational health and safety management systems – Requirements with guidance for use. ISO. https://www.iso.org/standard/63787.html.

2 Isniah, S., Purba, H. H., & Debora, F. (2020). Plan do check action (PDCA) method: literature review and research issues. *Jurnal Sistem dan Manajemen Industri*, 4(1), 72–81.

7 Design and Planning

Context drives behavior, and the context of work is shaped by decisions made long before employees step foot on-site. For this reason, EBS principles must be integrated as early as possible in a project or task, when our ability to influence outcomes is greatest.[1] Prevention through Design (PtD) provides a structured framework for this integration, allowing us to address high-energy hazards before they appear on site. Although design changes may require more resources upfront, it can still be much cheaper than facing the consequences on site.

Safety is often mischaracterized as a field activity, leaving front-line workers to navigate dangerous tasks, congested environments, and resource constraints. Research consistently shows that early intervention shifts safety management from reactive to proactive, addressing high-energy hazards at their source.[2] These efforts, though invisible to front-line workers, have a profound impact: fewer high-energy hazards to manage, smoother workflows, and safer worksite configurations.

This chapter explores how to integrate EBS into the design process. Here, "design" encompasses all decisions made before work begins, including final product design, worksite configuration and task sequencing. Each step emphasizes anticipating and mitigating high-energy hazards and SIF precursors in how work is imagined.

IMPLEMENTING EBS EARLY IN DESIGN AND PLANNING

In Chapter 4, we introduced the new hierarchy of energy control (HOEC), which prioritizes the elimination and reduction of energy. These energy controls are most effectively achieved through design decisions that directly eliminate or reduce energy exposure as shown in Figure 7.1. As work progresses from design to on-site planning, it becomes progressively more difficult to reduce or eliminate energy. By the time work reaches pre-job briefs and active execution, we are often limited to specialty PPE and Alternative Controls, which are more prone to human error. In short, it is more effective and efficient to mitigate energy before work begins.

There is an odd tension when it comes to PtD. On the one hand, many argue that it is the most effective way to prevent injuries.[3] It has been promoted by NIOSH and even institutionalized in the UK through the Construction Design and Management (CDM) regulations. Despite these efforts, PtD is still uncommon. In countries like the United States, the OSHA General Duty Clause[4] removes most responsibility from designers and planners. Instead, it puts the responsibility for safety on employers, such as contractors and subcontractors who hire front-line workers. This problem gets worse with typical hard bid, design-bid-build delivery strategies that divide design and construction processes and compress schedules. In contrast, collaborative contracting strategies, such as integrated project delivery, facilitate the alignment of design and planning efforts, ensuring safety is prioritized alongside cost and schedule considerations. Since the methods of project delivery and contract terms are

DOI: 10.1201/9781003641483-9

selected by clients, PtD is driven by their interests and decisions. Put simply, when it comes to PtD, clients play the pivotal role.

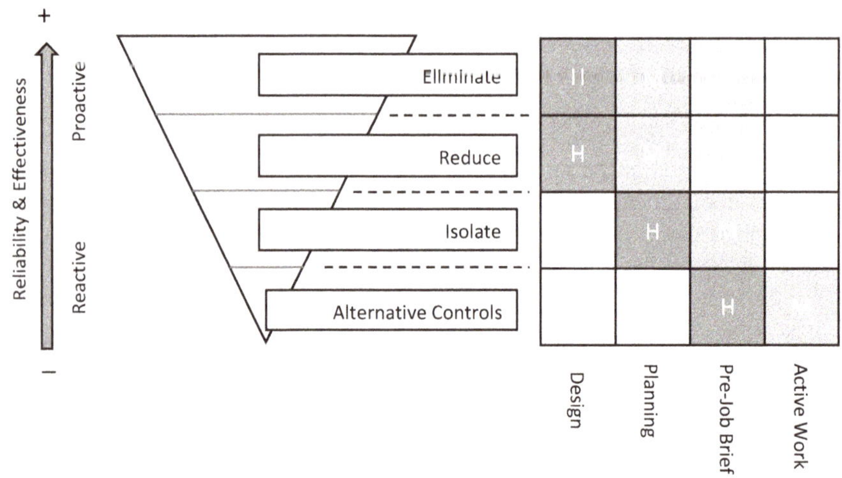

FIGURE 7.1 Energy control opportunities and relative effectiveness across stages of work delivery. Energy controls can be most effectively implemented across the work lifecycle. Opportunity levels (H = high, M = medium, L = low) are shown alongside a curve representing the declining impact of controls as work progresses.

EBS IN THE DESIGN OF THE FINAL WORK PRODUCT

The essence of PtD involves changing the features of the work product so it is safer to build, operate, and maintain. These "products" may range in scope and complexity from a simple tool or component, to a large capital project like a hospital. Although there are many ways to "do" PtD, we will focus on design charrettes because they are action-oriented and applicable in most industry sectors.

A design charrette is an intensive, collaborative session where stakeholders such as designers, engineers, safety professionals, contractors, and operators work together to consider constructability, efficiency, and safety. By integrating EBS principles into design reviews, we can efficiently identify and mitigate high-energy hazards as work is conceived. For example, I once participated in a charrette for a hospital construction project where permanent parapet walls were designed to eliminate the fall hazard across the project lifecycle. Similarly, for a manufacturing facility, the team redesigned the placement of high-voltage electrical panels and overhead crane pathways to reduce the risks of electrical arc flash and collisions, respectively.

SIF-FOCUSED DESIGN CHARRETTES

How do we address high-energy hazards before work begins? The following is a recommended structure for a SIF-focused design charrette. Ideally, safety would be

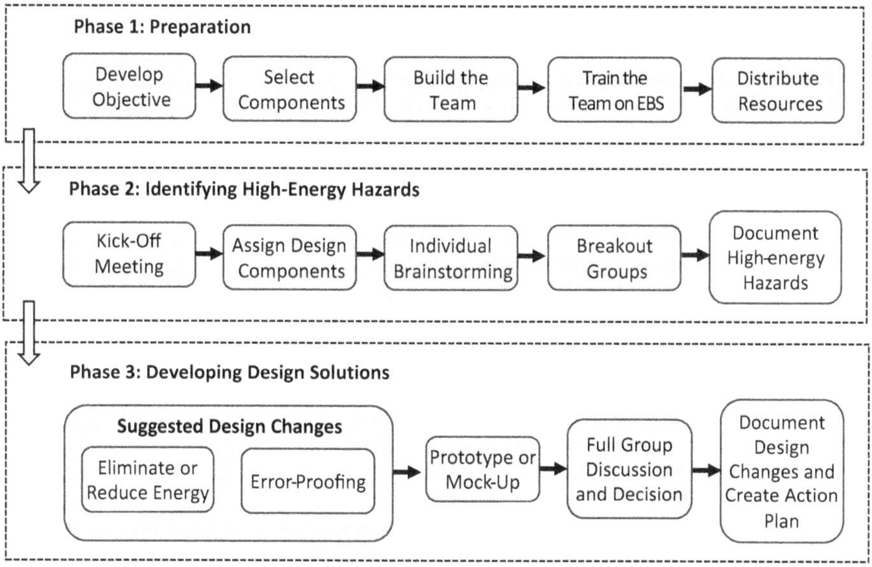

FIGURE 7.2 SIF-Prevention Design Charrette Process. The charrette is structured into three phases: preparation, identifying high-energy hazards, and developing design solutions to control them.

considered as an integrated component of an existing constructability review process. Although these reviews are common for large, sophisticated capital projects, this is not always the case. Therefore, this review will present a SIF-focused design charrette as a standalone process that can be implemented independently or integrated with existing processes.

Readers familiar with HOP may notice a similarity between a design charrette and a learning team.[5] In many ways, a design charrette is a learning team that is specifically centered on how high-energy can be mitigated in design. The process follows three integrated phases as shown in Figure 7.2 and each step is described below.

Phase 1: Preparing the Team

- Develop Objective: Before establishing the team and starting the charrette, it is important to define the purpose. For a SIF-focused review, the goal should be to: *Eliminate and reduce high-energy hazards during construction, operations, and maintenance by changing the features of the product.* This will serve as the North Star, helping the team to stay focused. Although all types of hazards may be considered, the primary focus will be on identifying high-energy hazards and developing practical design solutions.
- Select Components: In complex projects like hospitals or commercial buildings, there may be many design components to consider. Clearly, not all can be reviewed in one meeting. Instead, three to five high-priority design components

should be selected based on trends in SIFs, potential SIFs, and high-energy exposures.[6]

- Build the Team: Engage a diverse group of stakeholders who bring different perspectives and expertise to form the team. At a minimum include operators, supervisors, and front-line workers who understand the work; designers and engineers who understand technical design and applicable codes; project managers or regional leads who understand the logistics of the work; safety professionals who understand the required policies and procedures; and client representatives who ultimately must approve any changes. A team lead should be selected to facilitate the meetings. To be effective, company and project leadership must give this team the authority to make design decisions.
- Training the Team on EBS: The team needs a solid understanding of EBS before they start with a review. This includes the energy wheel, high-energy hazards, and types of controls. After the training, EBS resources such as the examples of high-energy hazards and Direct Controls in Appendix 3 should be shared.
- Distribute Resources: The facilitator should gather and distribute relevant background information about the project such as the current design, location and site conditions, financial and logistical constraints, and any other information that will help the team understand the status and setting of the design. The facilitator should also provide the results from previous design charrettes to help the team understand the assignment, inspire ideas, and avoid reinventing the wheel.

  ***Outcome of Phase 1*:** A diverse group of well-trained professionals who have the information and resources they need to eliminate or reduce high-energy hazards by changing the design.

Phase 2: Identifying High-Energy Hazards

Once assembled and informed, the team's first job is to review the design to identify as many high-energy hazards as possible.

- Kickoff Meeting: Begin with a session, so the team can get acquainted, review the goals, and develop a clear understanding of their role in the process.
- Assign Design Components: Break the team into sub-groups with each assigned one design component at a time. For example, if three design components will be reviewed in this design charrette, break the team into three evenly sized groups, assigning one component to each group. See Figure 7.3.
- Independent Brainstorming: Begin by asking team members to *individually* identify high-energy hazards associated with the assigned design component. Use the energy wheel and high-energy icons to identify as many high-energy hazards as possible. For example, the high-energy icons can be used to prompt specific questions, such as: Does this task require work near suspended loads, heavy mobile equipment, toxic chemicals, etc.? Include prompts for all 13 high-energy icons and instruct the team to consider construction, operation, and maintenance tasks in their review.

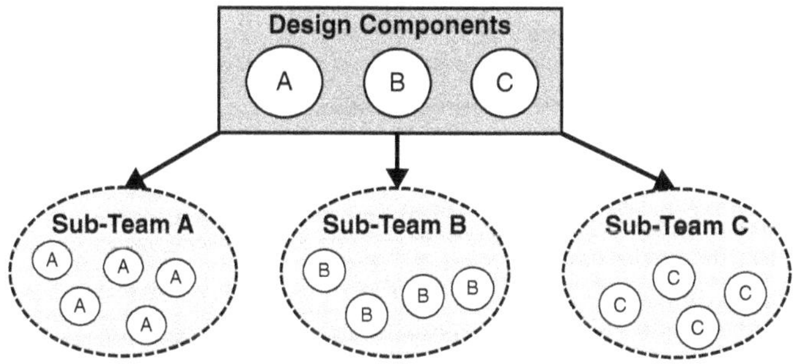

FIGURE 7.3 Assignment of design components. This figure illustrates how participants can be divided into sub-teams, with each group focused on a different design component during the charrette process.

- Breakout Groups: Convene the groups and ask each team member to share the hazards that they identified. Use a round robin approach until all identified high-energy hazards identified by the group members are documented.

  Outcome of Phase 2: Comprehensive catalog of expected high-energy hazards for each design component reviewed.

Phase 3: Developing Design Solutions

Once all high-energy hazards have been identified, the team can move toward developing solutions.

- Eliminating or Reducing High-Energy: Have breakout groups brainstorm how each high-energy hazard could be eliminated or reduced to a low-energy state. For example, a team may explore how an electrical panel could be lowered so a ladder is no longer required for installation or access, or how a mechanical system could be redesigned to be low voltage. In the literature, hundreds of design change examples have been documented[7] and many are integrated into digital tools such as the Design for Safety Toolbox and ToolSHeD™.[8]
- Error-Proofing the Work Environment: Clients who own and operate a facility may be interested in how the design influences the behavior of operators and maintenance workers. In addition to eliminating or reducing energy, the team can also look at ways to make the design simpler such as using standard parts and arranging features in a clear, easy-to-use layout.
- Prototypes or Mock-Ups: Not all design changes will be simple. Some design changes may be costly, complex, or introduce new risks. Physical or digital prototypes can help reveal those challenges early. This can be a great application of emerging technology such as 3D models, building information modeling (BIM), and virtual design and construction (VDC).
- Full Group Discussion and Decision: Since design changes can be expensive and complex, breakout groups should propose the change to the larger group. The full team of diverse stakeholders can ask questions, provide feedback, refine solutions, and reach a conclusion.

- Action Plan: Document the high-energy hazards in the original design, The design changes selected, and the rationale. Outline the steps for implementation, including responsibilities and timelines.

  Outcome of Phase 3: Design changes and action plans that eliminate or reduce as many high-energy hazards as feasible. This is also a great opportunity to reflect on the number of high-energy hazards removed or controlled through the design decisions (e.g., we started with 24 high-energy hazards, eliminated 6, reduced 4, and designed engineering controls for 2. Now, 12 remain that still must be addressed during design and operations).

  The processes and constraints can be different across sectors and geographies. Although this chapter presents a generic charrette, you may need to adapt the process or even repeat some steps as complex designs transition from conceptual design to detailed design documents.

EBS IN THE DESIGN OF WORK

Some hazards don't come from the product at all, they show up because of how the work is organized and planned. Although the client has final say on the design of the final product, contractors often control the worksite, the means and methods of their work, and how tasks are sequenced (see Figure 7.4). Unlike changes to the design of the final product, which require the input of many stakeholders, the construction team may have more freedom in the design of the work environment. Reviews must include input from those who perform the work, their immediate supervisors, and the managers who coordinate tasks and deploy resources.

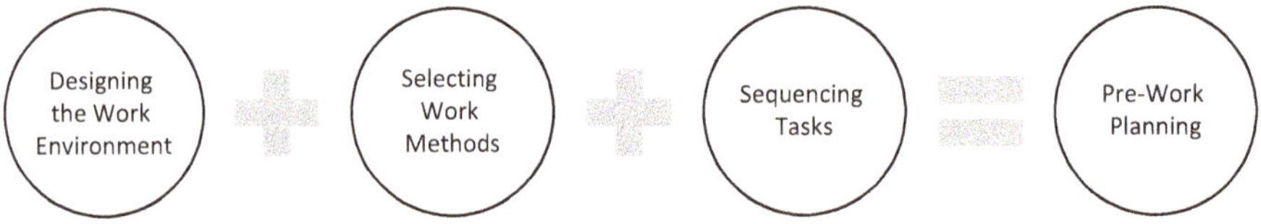

FIGURE 7.4 Three key elements of pre-work planning. By shaping the physical environment, work approach, and task sequence in advance, teams can proactively reduce exposure to high-energy hazards.

Physical Work Environment

The way materials, equipment, tools, and staging areas are organized affects exposure to high energy. For example, crane operations should be placed strategically with associated exclusion zones; there should be clear delineation and barriers between equipment pathways and pedestrian pathways; and electrical panels and circuits should be strategically located to minimize the need for workers to operate nearby. The worksite should be designed to be predictable and to isolate workers from high-energy hazards such as heavy machinery, electrical equipment, suspended loads, excavations, and holes.

The way the site is organized can also impact SIF precursors and human error. Sites that are cluttered and disorganized are much more error-prone than those that are orderly and logical. This is especially important when SIF precursors such as fatigue and distractions are present.

WHEN CONSIDERING WHAT THE WORKSITE WILL LOOK LIKE, ASK THE FOLLOWING QUESTIONS:

- What high-energy hazards are expected?
- Where will each high-energy hazard be located (consider both static locations and pathways for hazards that will be moving)?
- What Direct Controls can be installed as part of the design (e.g., barriers)?
- What Direct and Alternative Controls will be needed during the work?
- In what other ways can we ensure that the site layout is orderly, logical, and easy to navigate?

Work Methods

The choice of equipment, tools, and methods can directly impact the control of high energy. Therefore, high-energy hazards should also be considered in how the work will be performed. For example, when work is expected at height, consider using elevated work platforms with guardrails instead of ladders; ensure that concrete formwork and shoring systems are well maintained and properly engineered for the work; design ventilation systems for confined spaces; and avoid the need for manual handling of unstable materials.

Whenever possible, select equipment that has integrated controls and features that limit the impact of errors. For example, choose cranes that have automated brakes, collision detection systems, load limiting, and anti-sway systems; heavy equipment with automatic brakes and collision detection; and tools with built-in safeguards such as interlocks, deadman switches, and guards.

Obviously, safety cannot be the only factor considered. We must weigh the cost and schedule implications of the alternatives we consider. For example, the selection of elevated platforms or alternative concrete formwork systems may be prohibitively expensive, but they should at least be considered.

WHEN CONSIDERING HOW THE WORK WILL BE PERFORMED, ASK THE FOLLOWING QUESTIONS:

- What high-energy hazards are associated with the way we plan to do the work?
- Can we do the work differently to eliminate or reduce high-energy hazards?
- What Direct Controls can be designed into the system so that it is inherently safe (e.g., automated brakes, guardrails)?
- When the high-energy hazard must remain, what Direct and Alternative Controls are needed, and can they be incorporated into the site layout?

Schedule

The way we sequence our work also impacts high-energy hazards and SIF precursors. Understandably, schedules are usually optimized for production efficiency. We determine critical paths and use float to minimize resource demands and achieve the shortest possible completion time.[9] Too often, this results in a schedule that is so tight that there is constant pressure to rush work. Clients and contractors must consider the trade-offs: Is it more important to finish a day earlier if we could significantly reduce the potential for a SIF? Can we use float to avoid piling high-risk tasks on top of each other?

The sequence of tasks can even create a SIF exposure directly. For example, the schedule may require work on foot when heavy mobile equipment is traversing the same space; welding may be required adjacent to suspended loads; or utility work may be required adjacent to traffic at peak times. These potential conflicts are exceedingly difficult to control on site, but can be removed by avoiding clashes as the work is sequenced.[10]

In addition to exposures to high energy, the schedule can also impact SIF precursors. The schedule directly impacts productivity pressure, long hours, and fatigue and indirectly affects the quality of pre-job briefs and plans for change. When people are rushed, fatigued, and stressed, the likelihood of slips, lapses, mistakes, and violations skyrockets. The schedule, and by extension the contract arrangement, is a contributor to SIFs when it causes abnormal pressure to work quickly.[11] Managing the schedule is a powerful but often overlooked way to prevent SIFs. To do it well, companies should train schedulers in EBS and include safety professionals in planning.

WHEN CONSIDERING THE WORK SCHEDULE, ASK THE FOLLOWING QUESTIONS:

- What exposures to high-energy hazards result from the way tasks are sequenced?
- When will those high-energy hazards emerge during the workday?
- How can we rearrange the schedule to eliminate or reduce high-energy exposures?
- How will the schedule impact SIF precursors?
- How can we use float to decrease work pressure?
- How is SIF prevention weighed against cost, quality, and productivity?

CONCLUSION

There is a common delusion that we can change safety culture through a few strategic actions. We can't. Real change takes time and comes from steady, focused improvements in how we do the work. The best way to initiate this change is by considering safety in design and planning. When companies make intentional decisions to prioritize safety, it demonstrates genuine care that is inevitably reciprocated among employees. Although changes in design and planning can be a challenge, they are far easier than dealing with the consequences in the field.

Client support for these initiatives is crucial to success. Without client commitment, EBS in design and planning remains an uphill battle. Clients must prioritize SIF prevention, even if it means compromising on schedule and budget. Although PtD can be costly, safety decisions can pay off throughout the lifecycle of a product through fewer injuries, claims, and disruptions.[12]

EBS provides a pathway to make safety in design and planning more focused and efficient. By aiming to eliminate and reduce high-energy hazards, we can concentrate on changes with the highest impact. EBS shifts our approach from "safety in design and planning" to "SIF prevention through design and planning," helping the team rally around the clear and compelling purpose of saving lives.

Strong design gives front-line workers their best shot at success. They often bear the brunt of poor design, aggressive timelines, and tight budgets. However, even within these constraints, they can still integrate EBS in their planning. High-energy hazards, Direct Controls, and SIF precursors can be considered in the organization of the site, selection of equipment, and scheduling. This approach helps make the worksite orderly, logical, and calm. *This is context by design.*

This chapter has reviewed the integration of EBS before work begins. In the next chapter, we will explore how EBS principles can be incorporated into daily site safety planning once work is underway.

KEY TAKEAWAYS

1. Preventing SIFs starts with eliminating and reducing high-energy hazards during design.
2. Design charrettes offer a structured approach to integrate EBS principles into the final product.
3. Clients play a pivotal role in SIF prevention by influencing project delivery, contract terms, schedule, and budget.
4. Contractors can address energy exposure and SIF precursors in worksite design and scheduling.
5. EBS makes safety in design and planning more focused and efficient.

NOTES

1 Schulte, P. A., Rinehart, R., Okun, A., Geraci, C. L., & Heidel, D. S. (2008). National prevention through design (PtD) initiative. *Journal of Safety Research*, 39(2), 115–121.

2 Toole, T. M., & Gambatese, J. (2008). The trajectories of prevention through design in construction. *Journal of Safety Research*, 39(2), 225–230.

3 Tymvios, N., Hardison, D., Behm, M., Hallowell, M., & Gambatese, J. (2020). Revisiting Lorent and Szymberski: Evaluating how research in prevention through design is interpreted and cited. *Safety Science*, 131, 104927.

4 Occupational Safety and Health Administration. (1970). General duty clause, Section 5(a)(1) of the Occupational Safety and Health Act of 1970. U.S. Department of Labor. https://www.osha.gov/laws-regs/oshact/section5-duties

5 Edwards, B., & Baker, A. (2020). *Bob's Guide to Operational Learning: How to Think Like a Human and Organizational Performance (HOP) coach*. Independently published.

6 This will be revied in Chapters 11 and 12. The EBS cycle begins with design and planning and ends with feedback to design and planning in the form of trends in incidents and observations.

7 Gambatese, J. A., Hinze, J. W., & Haas, C. T. (1997). Tool to design for construction worker safety. *Journal of Architectural Engineering*, 3(1), 32–41.
8 Cooke, T., Lingard, H., Blismas, N., & Stranieri, A. (2008). ToolSHeDTM: the development and evaluation of a decision support tool for health and safety in construction design. *Engineering, Construction and Architectural Management*, 15(4), 336–351.
9 The critical path is the sequence of tasks that determines the shortest possible duration to complete work, where any delay in these tasks will directly delay the overall completion. Float in the schedule refers to the amount of time that each task can be delayed without affecting the overall completion time.
10 Esmaeili, B., & Hallowell, M. (2013). Integration of safety risk data with highway construction schedules. *Construction Management and Economics*, 31(6), 528–541.
11 Hallowell, M. R., Bhandari, S., & Alruqi, W. (2020). Methods of safety prediction: Analysis and integration of risk assessment, leading indicators, precursor analysis, and safety climate. *Construction Management and Economics*, 38(4), 308–321.
12 Alkaissy, M., Arashpour, M., Li, H., Alaghmand, S., & Nezamian, A. (2022). Quantitative analysis of safety risks and relationship with delayed project completion times. *Risk Analysis*, 42(3), 580–591.

8 Pre-job Safety Briefs and Management of Last-Minute Change

If your company could be amazing at only one safety activity, what would it be? I have asked this question of countless leaders as we explore their priorities and where their resources should be spent. The answer is almost always the same: pre-job safety briefs (PJSBs).[1] At its core, a PJSB is a safety-focused discussion that takes place on site immediately before work begins. The objective of this activity is to break the work into individual tasks, identify potential hazards and necessary safeguards, and define roles and responsibilities. Generally, mature companies expect every crew to have a PJSB before each work period, after significant breaks, and anytime work is stopped.

PJSBs are the cornerstone of most safety programs and, according to research, they drive safety performance more than any other tactic.[2] When performed poorly, inadequate PJSBs have been shown to be one of the strongest precursors of SIFs. When performed well, PJSBs ensure that all team members understand their tasks, potential hazards, and safety procedures, significantly reducing the risk of SIFs.

Companies make *massive* investments in their PJSBs, sometimes without even realizing it. I once did the math to show a large contractor the magnitude. Given the parameters shown below, we determined that they invested over $7,000,000 USD per year just on employee time spent on PJSBs. When accounting for replanning after breaks, paperwork, leadership presence, or supporting systems, the estimate was closer to $10,000,000 USD per year. These numbers are surprising to most because PJSBs are often seen as part of the workday rather than as a significant investment. I highly recommend replacing this example assessment with your own numbers; it can make a compelling case with operations and senior leaders.

- 275 crews and an average of 4 employees per crew
- 300 working days per year
- 15 minutes spent holding one PJSB per day
- $90 per hour loaded labor rate for the skilled workers

Clearly, if companies are going to commit such substantial resources, PJSBs better be done really well! When it is just a check-the-box exercise, it is a missed opportunity and an incredible waste of time that could otherwise be spent on productive work.

DOI: 10.1201/9781003641483-10

Alternatively, research shows that a quality PJSB that integrates the principles of EBS will pay dividends.[3]

This chapter reviews the attributes of a high-quality PJSB and explores how to incorporate the EBS principles to emphasize SIF prevention. Although integrating high-energy and Direct Controls into a PJSB may seem straightforward, the key to success lies in embedding the topics within a robust and engaging discussion. EBS has a much greater potential for success when there is engagement among all employees, high competency, strong leadership, and genuine care. This is the fertile soil that makes EBS thrive.

In addition to pre-work planning, we will also explore how EBS principles help crews manage last-minute change during work. Although it would be easy to tell employees to have an excellent PJSB and then follow their plan, work is never that simple. Unanticipated changes can arise, and if left unaddressed, some may lead to catastrophic outcomes. By integrating EBS with PJSB and Stop Work Authority, we will have a plan for the work as it's envisioned, and a process for adapting when it changes.

CASE STUDY OF A WEAK PRE-JOB SAFETY BRIEF

I have had the opportunity to observe hundreds of PJSBs across companies, sectors, and geographies. My biggest surprise was the huge variation in quality, even within the same company and on the same project. Some are fantastic because everyone speaks up, takes the discussion seriously, and does their best to identify and mitigate hazards. Others are merely a "meeting" where some crew members complete paperwork in silence while others kick rocks in the background. To ensure high-quality discussions, we must understand and communicate what "good" looks like.

Years ago, we were researching the state-of-the-art in PJSBs and were given permission to record a meeting. The meeting was perfectly average, and it serves as a nice starting point for our discussion about quality. Below, a description of the work and a transcript of the meeting are provided as a case study. Give it a read and make your judgment. What did this group do well, and how could they improve?

WORK DESCRIPTION

Where: The meeting is taking place on the ground floor of a commercial building project.

Who: Three crews with four workers each. The crews include electricians, plumbers, and painters.

What: Each crew is performing their respective tasks on the eighth story of the building.

When: Work hours are longer than usual because the team is working toward the certificate of occupancy.

How: Typical work tools for the trades and typical installation practices will be used.

MEETING TRANSCRIPT

Crew Lead: Good morning, everyone. It feels like a Monday, that's for sure. It rained last night, so I'll be out cutting the grass tomorrow. Anyone else?

Alright, everyone, listen up.

Let's focus for these next eight hours. Keep an eye on each other and make sure we get the job done safely. Check those extension cords—look for any cuts or abrasions. If there is a problem with the tools, let us know.

Also, pay attention to the ladders. Make sure they're stable before you use 'em. If you see any spills or hazards, clean up or let someone know.

Note: There was no commentary from the other 12 crew members.

Alright, that's about it. If anyone has anything, now's the time to speak up. Otherwise, let's get to work.

Was this a perfect PJSB? No, not by a long shot. However, it's far from the worst meeting. It at least fulfills the basic requirements: on site, conducted immediately before work, focused on safety, everyone was present, and the paperwork was detailed and complete. The planning process wasn't terrible, but it could have been much, much better.

Let's be clear, the leader wasn't incompetent, nor did he set out to hold a subpar meeting. Quite the opposite. He was an experienced electrician who had recently moved into a site leadership role. Like most, he conducted the meetings in the same manner he had observed for years, perpetuating both good and bad practices of the leaders who came before him. He had the background to lead a great PJSB – he was respected, knowledgeable about the work, and cared – but he had never been shown what good looks like. If we want to break this cycle, we must equip field leaders like him with training on how to facilitate an excellent PJSB.

The key attributes of an excellent PJSB have been recognized for ages but remained hidden in pockets of undocumented knowledge within the profession. In one of our research projects, we aimed to unearth this valuable wisdom, distill it to essential features, and develop resources to empower those leading these critical safety meetings.

FEATURES OF AN EXCELLENT PJSB

We recognize a good meeting when we see it, but quality is hard to define without clearly stating the exact features we want to see. Precise communication of carefully chosen features transforms a vague idea of quality into a tangible reality. By being clear about our expectations, we can make quality something we can see, measure, and achieve.

In one of the first projects undertaken by the CSRA, a team of 23 safety professionals with a collective 300+ years of industry experience was formed to document the features of a high-quality PJSB. Through brainstorming sessions, the team identified over 50 features of a good meeting, and the list was refined to the 15 most important elements using the Delphi method.[4] The key discussion points for each of these elements were summarized, collated, and published in a field guide. EBS was a key theme, where excellence included using the energy wheel to ensure that hazard planning was complete and giving extra attention to the identification and control of high-energy hazards.

In the next section, we will review the highlights of the team's output with emphasis on integration of EBS. The guidance is focused on what a person leading the meeting such as the foreman or crew lead (i.e., the "facilitator") can do before, during, and after the PJSB. While the quality of the meeting reflects broad participation, the facilitator plays a pivotal role in ensuring its success.

Before the Meeting

An excellent PJSB begins before the meeting ever starts. In preparation for the meeting, the facilitator should gather enough information so they:

- **Know the task** including the sequence, configuration, materials, tools, equipment, and other attributes of the work.
- **Know the environment** such as the different crews that may be sharing the workspace, weather, or terrain.
- **Know the schedule** of activities that may be performed at or near the workspace.
- **Know the situation** by identifying anything out of the ordinary or any changes or disruptions that may occur.
- **Know the history** of the crew including levels of experience, any recent safety-related challenges, successes, or failures.
- **Know the principles of EBS** including how to use the energy wheel, the high-energy icons, and how to apply the rules of Direct and Alternative Controls.
- **Know the crew** so that precursors such as fatigue and distraction can be identified.

It's easy to tell when a facilitator has done their homework. They describe the task and expectations clearly, precisely, and with authority, setting the tone for the meeting and demonstrating its importance. A facilitator knowledgeable in EBS, equipped with tools like the energy wheel, high-energy icons, and definitions of Direct and Alternative Controls, can effectively help crew members learn and apply these principles. This preparation by the facilitator should typically take place the day before the task is planned.

During the Meeting

Once the pre-job safety meeting is set up for success, a meaningful discussion can take place with the active participation of all employees. The following 15 features, identified in the research, support a quality planning exercise that enables EBS to thrive.

1. ***Ensure that everyone performing the job is present.*** All crew members working on the job must be physically and mentally present at the meeting. That means being engaged in the process and free of distractions like cell phones. When a crew member is absent for the discussion about high-energy hazards and controls, it can result in a critical gap in knowledge and assumptions that could lead to SIFs.
2. ***Ensure that everyone actively participates.*** An excellent facilitator ensures active participation by asking questions, seeking clarifications, and recognizing the value of workers' ideas. To achieve this, the facilitator should use open-ended questions, involve shy or less experienced workers, and acknowledge each worker's contributions, fostering a sense of ownership and value. Instead of merely completing a checklist or reading paperwork, the facilitator should cultivate a vibrant discussion where all workers contribute.
3. ***Hold the discussion as close to the work as reasonably possible.*** Holding the discussion close to the work area allows the crew to point out and inspect specific high-energy hazards and avoid assumptions or misunderstandings. When the meeting must be held elsewhere, a re-huddle should take place before work begins to discuss any high-energy hazards and associated controls that may have been missed.
4. ***Break down the task into individual work steps.*** As the discussion begins, the crew should break down their planned tasks into major steps. They should also establish clear roles and responsibilities, and ensure everyone has appropriate training and competence for their assigned roles.
5. ***Ensure the necessary tools and equipment are available.*** The crew should discuss the resources they need to perform the job safely and efficiently. This helps reduce the need for improvising or working around high-energy hazards without the necessary gear. Understanding the task steps, roles, tools, and equipment provides the foundational information needed to begin anticipating hazards. If the correct tools or equipment are not available, the facilitator should discuss the issue with management.
6. ***Identify hazards associated with each task.*** Perhaps the most important activity in a PJSB is the identification, prioritization, and control of hazards, especially those that are likely to cause SIFs. This begins by identifying hazards that are inherent to each task such as a suspended load required to position equipment or work at height required to paint. Safety can sometimes be reduced to soundbites like "slips and trips," "pinch points" and "line of fire." These generic hazards should always be further clarified. Although it can be tempting for a knowledgeable facilitator to identify hazards *for the workers*, hazards should be identified and discussed *by the workers*.

   This step is the true beginning to EBS integration in a PJSB. To ensure that the hazard scan is broad and complete, the energy wheel should be used as a tool to remind workers of hazard types that are commonly overlooked. To start, the crew should identify all hazards that come to mind. Then, after ideas are exhausted, the energy wheel can be used as a prompt to find hazards that were overlooked in the initial pass. The facilitator should wait until workers have finished identifying hazards on their own before reviewing any that were missed. The energy wheel helps ensure a broad scan that includes both low and high-energy hazards. In Step 9, we will ensure a focus and a proportional response to high-energy hazards.

7. ***Identify hazards in the surrounding work environment.***
   Many times, the scan for hazards stops with the tasks. However, in most work, hazards also stem from the surrounding environment. For example, there may be factors related to the weather, terrain, influence of the public, and other crews working nearby. These hazards are sometimes significant and challenging to control. Following the same protocol discussed previously, hazards in the surroundings can be identified using the crew's experience, the energy wheel, and guidance from the facilitator.
8. ***Discuss how each hazard will be addressed.***
   No matter the severity, every hazard must be mitigated. For example, there may be protocols to deal with insect bites, uneven walking surfaces, dust, and exposure to the sun. Once all hazards have been addressed, special treatment should be given to high-energy hazards.
9. ***Identify high-energy hazards and their controls.***
   Specific time should be allotted to identify and plan for high-energy hazards. With a full list of hazards from the previous steps, this discussion should give special consideration to the subset that exceed 1500 Joules. This is where SIF prevention comes into focus. To help with this step, the crew may wish to reference the high-energy icons from Chapter 4 and the high-energy and Direct Control examples in Appendix 3.

   Once all high-energy hazards are identified, the crew should ensure that a Direct Control is available for every high-energy hazard. As a reminder, Direct Controls must be targeted to the high energy; mitigate energy when installed, verified, and used properly; and work even when there is unintentional human error during the work. Since not all Direct Controls are equal, the crew should strive to eliminate, reduce, or isolate the high-energy. For example, when attempting to protect against an arc flash during switchgear maintenance, the crew may consider de-energizing the electrical system, installing a physical barrier around live electrical units, or wearing arc-rated gear, in that order.

   If a Direct Control is not feasible, an Alternative Control system should be established. Alternative Control systems are less favorable because they require multiple coordinated layers and are vulnerable to human error (see Chapter 6). However, an Alternative Control system has been deemed sufficient when a Direct Control isn't possible.

   Like the other discussions, the facilitator should encourage the workers to identify controls and step in when one is missed, if a higher order control is available, or if the plan requires further clarification.
10. ***Obtain and review required permits.***
   For many high-energy hazards, the client will require permits. This may include confined space entry, electrical, excavation, lifting, and hot work permits.

Although not a control in their own right, permits provide further documentation for high-energy hazards and clarify which controls are required to be in place during work. This redundancy underscores the importance of these high-energy hazards and reinforces the discussion held by the crew. As each high-energy hazard is discussed, the associated permit should be obtained and reviewed.

11. ***Create a plan to address change.***
Work rarely goes exactly as planned and change is a common root cause of SIFs. The crew should create a plan for change to ensure they are equipped to respond appropriately. The facilitator may consider the following prompts: *What might change today? How does the potential change impact our safety? What will we do if an unexpected change occurs?* Planning for change can be a challenge because there are seemingly infinite changes that could occur. Later in this chapter, we will review how to integrate EBS during work to manage change and apply Stop Work Authority.
12. ***Discuss the importance of stopping work.***
Stop Work Authority is the right and responsibility of any worker to halt work immediately if they identify a hazardous condition that could potentially lead to a SIF. This proactive measure empowers employees to address SIF exposures before resuming work. During the PJSB, the facilitator should remind the team that they have the authority and duty to stop and replan if they find themselves exposed to an uncontrolled high-energy hazard. To avoid confusion and increase worker confidence, the team should discuss the types of conditions when work should be stopped and how they will communicate. Along with planning for change, Stop Work Authority will be discussed in detail later in this chapter.
13. ***Review emergency response plans.***
A critical component of SIF prevention is having a plan in case of an emergency. An effective response plan includes emergency contact information, defined roles and responsibilities, detailed procedures, communication plans, equipment, and post-emergency procedures, among others. These plans are critical for SIF prevention because they ensure swift response and provide coordination and communication so that incidents do not result in more severe outcomes.
14. ***Document the PJSB.***
In our team, we intentionally prioritized meaningful discussion over paperwork, arguing that the main goal of a PJSB is to have an excellent discussion, not just to complete forms. It's unfortunate how often employees are conditioned to treat a PJSB as a paperwork exercise when, in fact, the discussion is the real prize. Although paperwork is not the purpose of the PJSB, it is an important output for verification and is required by some regulators. Absent a recording of the meeting, the documentation is the next best evidence that the meeting was held, and high-energy hazards and their controls were considered. At a minimum, documentation should include the tasks, hazards, safe work practices, with special designation of high-energy hazards and their controls. The final document should be acknowledged and

signed by the entire team, indicating that everyone understands, approves, and will follow the plan. This seemingly minor gesture can generate a sense of ownership of the plan.

15. ***Demonstrate understanding***.
    Three-way communication, or repeat-back communication, is crucial for SIF prevention. When the facilitator asks the crew to describe the safe work plan in their own words, it helps confirm the completeness of their understanding. Without such a check, there is a higher risk of miscommunication and misunderstandings, which can lead to uncontrolled energy even when a good plan is in place.

In my experience, these steps work well when work is dynamic, but for crews performing the same tasks daily, PJSBs can become repetitive, leading to disengagement. While covering usual hazards is important, shifting part of the discussion to what might be different today, such as weather, equipment condition, or nearby activities, makes the meetings valuable. The same EBS principles apply, but with greater emphasis on anything out of the ordinary. This approach helps prevent complacency and ensures routine work receives the same level of scrutiny as higher-risk tasks.

CASE STUDY OF A STRONG PRE-JOB SAFETY BRIEF

The first case study provided an example of a meeting that met the basic requirements but was devoid of the attributes of a high-quality PJSB. The following transcript was designed to show a meeting that is outstanding, but not perfect. Take a moment to review. What did they do well and how might they improve? How does this compare to the first case study?

WORK DESCRIPTION

Where: A residential urban setting at the end of an alley accessible to private vehicles. The meeting takes place next to the pole where work will be completed.

Who: An electrical utility crew with one crew lead and two employees.

What: The task is to replace a 120 lb recloser on the overhead electrical pole. The top of the recloser sits approximately 2 ft 6 inches below the primary electrical lines and approximately 25 ft above the ground.

When: Work is scheduled during regular work hours.

How: Work will be performed aloft in the bucket of a bucket truck.

Notes: A recloser provides overcurrent protection through fault isolation and automatic restoration for temporary faults on overhead sub-transmission lines. Operations Control Center (OCC) clearance is a required part of the work permit.

TRANSCRIPT OF THE PRE-JOB SAFETY BRIEF DISCUSSION

Crew Lead: Alright, everybody here?

Crew Member: Yes, sir.

Crew Lead: Today's job is to replace the recloser. We're gonna set up the truck in the right position. I'm gonna call the OCC, get clearance permit, and then I'll be on the ground to monitor you and have whatever tools and equipment you need.

Employee 1: I'll grab the extendo stick for pulling out the fuses, the gri-pall for the taps, rigging for that recloser to hang it, and some impact tools.

Crew Lead: OK, How about protection?

Employee 2: For protection, we're going to use our gloves and sleeves, and we'll also get the rigid barrier, the tubes to go up on the hotlines. We're gonna come down, we're gonna take it off the palette, get all the connections prepped and get it ready to go so it's easier to land. When we get ready to land the secondary closure, we will make sure we mount it right, and all connections are good.

Crew Lead: OK, when we get ready to land the second recloser, we need to make sure we mount it right and the connections are all good. Then, I'll call OCC, get permission to re-energize.

Alright, let's talk about high-energy. What do you see today and how are we work around them?

Employee 1: We got electrical work, so we're going to put up our hoses, and we will use our gloves and sleeves. We got our suspended load, and we need to make sure that all our rigging secure, and that nobody walks underneath it.

Employee 2: We're up in the bucket, so we need to make sure we're harnessed and buckled in. And then while we're flying the switch out to land it, we'll make sure that the rigging is all good to go, and our line of fire kit set up, lights all set up and everything.

Crew Lead: Anything else around us?

Employee 1: Let's block the alley with cones. As we're doing that, we will make sure all gates and fences are closed so dogs can't get out.

Employee 2: We will make sure there's no traffic in the area, put out some cones, make sure cars don't go driving through underneath us. And then make sure that we do have that line of fire kit set up.

Crew lead: OK, lastly, just watch out for tripping hazards while walking. You know, uneven surfaces back here. OK everyone clear on their assignments and the hazard controls?

Employees: Yup.

In our review, we found the following positive attributes of the PJSB:

- Everyone performing the job was present
- Everyone actively participated
- The discussion was held as close to the work as reasonably possible
- The task was broken into individual work steps
- Necessary tools and equipment were discussed
- Hazards associated with each task were identified
- Hazards in the surrounding work environment were identified
- Mitigations were identified for each hazard
- High-energy hazards and their controls were emphasized
- Permits and required approvals were discussed
- The PJSB was documented
- Crew understanding was confirmed

Some areas where this crew could have improved are as follows:

- Plan to address change
- When and how to use Stop Work Authority
- Review of emergency response plans

Individual assessments of PJSB quality can vary because everyone has at least slightly different expectations. However, the one area where the crew clearly needed improvement was planning for change. Even well-trained crews have difficulty in this area because we provide them with very little guidance. In the next section, we will explore how EBS principles can guide the management of last-minute change during work.

USING EBS TO ADDRESS LAST-MINUTE CHANGE

One of my pet peeves is when workers are instructed by management to "plan for change and stop when something changes" because change is a regular part of most work. By the end of the workday, there can be a significant disparity between how the crew envisioned the work and how it actually unfolds. As we learned, about 30% of all hazards emerge during work and cannot be anticipated during planning. Therefore, we need a robust plan for what to do when something does change, and guidance on when change is significant enough that it demands that work should be stopped.

The truth is, most managers don't define what "change" means, and they don't really know when they want their workers to stop and replan. Instead, they rely on unclear directions that ultimately put this uncertainty onto the shoulders of the workers. It isn't until an incident occurs that management provides feedback, typically in the form of a conclusion that the crew should have followed the safe work plan or stopped to replan. This is counterfactual reasoning[5] and an absurd conclusion. We can do better. The EBS principles and the focus on SIF helps us address this issue objectively.

Defining "Change"

A change is the instance of becoming different.[6] There is no specified threshold for when change becomes relevant or significant; this depends on the context and interpretation. Given that nearly 80% of SIFs are related to some form of change, we must use this term more intentionally and precisely.[7] In the context of SIF prevention, consider the following definition of a relevant change.

> ***SIF-Related Change*****:** An unexpected condition or action that occurs during work and results in an exposure to a high-energy hazard.

While the general word change is so broad that it is meaningless in practice, the definition of SIF-related change helps us begin to understand when changes are significant enough to require action. It helps us reframe direction from "plan for change" to "plan for what to do when there is an unexpected condition or action that occurs during work and results in an exposure to a high-energy hazard." This moves us from vague direction to something that is meaningful and actionable.

When to Use Stop Work Authority

Like pre-job safety briefs, Stop Work Authority also goes by many different names such as all stop; safety stop; stand down; and see something, say something. As we discussed, Stop Work Authority is the general term for the official right and responsibility to stop work when a perceived unsafe condition is recognized. If we frame this within the definition of a SIF-related change, we can define Stop Work Authority as the formalized right and obligation to pause and replan when there is an unexpected high-energy exposure. In other words, we should exercise Stop Work Authority when there is SIF-related change. Below is some guidance on using EBS to help determine when to apply Stop Work Authority.[8]

Stop work if any of the situations below are true for a high-energy hazard:

- A new hazard has emerged
- We identified a hazard that we missed during planning
- A hazardous condition has crept from a low to a high-energy state
- The location of the hazard and its proximity to workers has changed
- An assumption made about the hazard is no longer true
- There is something about the hazard that we don't understand

Stop work if any of the situations below are true for a Direct or Alternative Control:

- We notice that a control is missing for an identified hazard
- A control is present, but it's not adequate.
- The control is present, but it's not being used properly
- Condition has shifted so the installed control is no longer effective

- An assumption made about the control is found to be untrue
- There is something about the hazard that we don't understand

Not every change requires a formal application of Stop Work Authority. For most low-energy changes, stopping and replanning is excessive, and a brief pause for the crew to communicate and resolve the issue is usually sufficient. Similarly, not all changes involving high-energy hazards are significant. Minor adjustments, such as moving a control, replacing barriers, or raising a guardrail, require communication and monitoring but may not necessitate stopping work. However, if changes cause significant exposure to high-energy hazards, work should be stopped, and a new plan should be created. This shows why monitoring high-energy hazards and Direct Controls is important. The relationship between the magnitude of change and the magnitude of energy is illustrated in Figure 8.1.

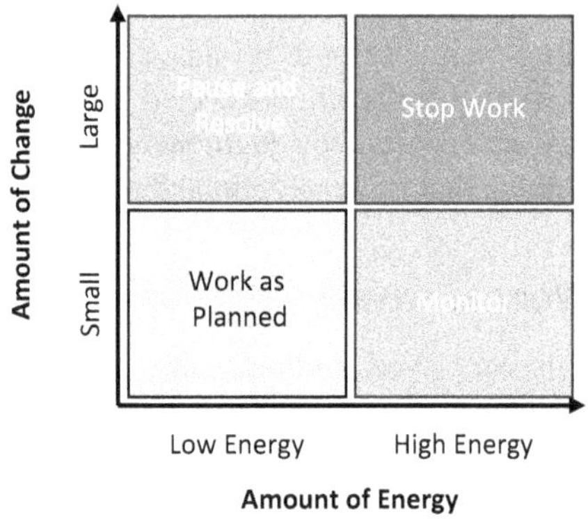

FIGURE 8.1 Guidance on applying Stop Work Authority based on energy and change. This decision aid helps teams determine when to pause work by evaluating the magnitude of energy involved and the extent of change. High-energy tasks with significant changes warrant the greatest caution.

CONCLUSION

As Vince Lombardi once said, "Perfection is not attainable, but if we chase perfection, we can catch excellence." This sentiment perfectly applies to PJSBs. While we may never achieve the perfect PJSB, striving for perfection will lead us to excellence in what is arguably the most critical daily safety activity.

When PJSBs are performed well, they create an ideal environment for EBS to succeed. The energy wheel helps crews scan broadly for hazards, the concept of high-energy focuses attention on the most significant risks, and verification of controls ensures confidence that everyone will go home safely at the end of the day. Research shows that training field leaders on conducting PJSBs directly enhances the quality of these meetings and the presence of controls during the work.

With a thoughtful design and an excellent plan in place, we must then turn our SIF prevention efforts to managing the realities of dynamic work. As conditions change, we can directly apply EBS principles to evaluate change and understand when application of Stop Work Authority is warranted. Not only does EBS help us to view this under the lens of SIF prevention, but it also turns a vague idea into something useful.

As safety efforts move from design and planning to the management of ongoing work, EBS begins to tie the safety program together. This is the all-important "red thread" of SIF prevention. In the coming chapters, we will examine how EBS adds further cohesion by guiding Safety Walks, measurement, and learning.

KEY TAKEAWAYS

1. Pre-Job Safety Briefs are a cornerstone of an effective safety program.
2. Focusing on the control of high-energy during pre-job discussions is critical to SIF prevention.
3. The characteristics of an excellent pre-job brief are well documented and can be used for training to define "what good looks like."
4. Unanticipated change during work contributes significantly to SIFs; crews must plan for change and implement robust processes to respond effectively.
5. For minor changes, crews should pause, pivot, and monitor, but uncontrolled high-energy hazards should trigger the crew to stop and replan.

NOTES

1 This activity goes by many names including toolbox talks, tailgate talks, pre-task plans, and field-level hazard assessments, and more.

2 Alruqi, W. M., & Hallowell, M. R. (2019). Critical success factors for construction safety: Review and meta-analysis of safety leading indicators. *Journal of Construction Engineering and Management*, 145(3), 04019005.

3 Bayona, A., Hallowell, M. R., & Bhandari, S. (2024). The Things That Hurt People Are Not the Same as the Things That Kill People: Key Differences in the Proximal Causes of Low-and High-Severity Construction Injuries. *Journal of Construction Engineering and Management*, 150(8), 04024088.

4 Note that the team's focus was on crews that spoke the same language. Lone-worker situations or those where one or more crew members do not understand the language spoken pose unique challenges.

5 Hoch, S. J. (1985). Counterfactual reasoning and accuracy in predicting personal events. *Journal of Experimental Psychology: Learning, Memory, and Cognition*, 11(4), 718.

6 Oxford University Press. (n.d.). *Oxford English Dictionary*. Oxford University Press. https://www.oed.com

7 Lee, W. H., Gambatese, J., Hallowell, M. R., & Nnaji, C. Impacts of Last-Minute Changes on Construction Safety. In *Construction Research Congress* 2024 (pp. 497–507).

8 We should always err on the site of caution. Therefore, we should also Stop Work if there is uncertainty or a suspicion that there may be imminent danger.

9 Safety Walks and Leadership Engagements

So far, we have discussed the integration of EBS principles into design, planning, and work execution, spanning the Plan and Do parts of the plan–do–check–act (PDCA) cycle. Even when these activities are executed well, there can still be gaps in the control of high energy stemming from overlooked hazards, insufficient safeguards, and inevitable errors. To address these gaps, companies need a robust process to verify the adequate control of high energy during work and ensure employees have the knowledge, resources, and support they need to succeed. These checks enable systematic learning from normal work that reveal strengths and gaps in SIF prevention efforts.

Safety Walks bridge the gap between planning and execution by providing real-time insights. Although this activity is known by many names, such as observations, audits, and inspections, the fundamental process remains the same: visiting a site, observing conditions, and engaging with people. Like safety in design and planning, research has shown that a well-executed Safety Walk is a key driver of safety performance.[1] Safety Walks are not simply symbolic "safety PR" activities; they are deliberate, technical exercises requiring preparation, expertise, and a focus on controlling high-energy hazards.

Integrating EBS principles into a Safety Walk strengthens its focus on SIF prevention. A SIF-focused Safety Walk is a structured process in which a leader (1) tours an active work site, (2) observes conditions, (3) engages with employees, (4) assesses the control of high-energy hazards and presence of SIF precursors, and (5) provides feedback and support (see Figure 9.1). Within the EBS framework, a Safety Walk becomes an opportunity to conduct both an Energy-Based Observation (EBO) of high-energy hazards and their controls, and a Field Safety Engagement to assess whether precursors are present.

Safety Walks are typically conducted by individuals in positions of leadership. In this chapter, we use the term "leader" broadly to refer to anyone with formal responsibility or influence who is not directly involved in the daily work of the crew they are visiting. This includes on-site roles such as foremen and site supervisors, and off-site roles like regional managers, executives, and safety professionals. The only individuals excluded from this definition are the crew's direct supervisor and their peers. That distinction is intentional. A Safety Walk is a form of independent checking, and for it to be effective, it must be led by someone with enough distance from the task to evaluate conditions and controls objectively.

Throughout this chapter, organizations that use a different label such as "observer" for those conducting Safety Walks may substitute terminology as needed. Regardless of the title, the principles and practices described here remain the same.

DOI: 10.1201/9781003641483-11

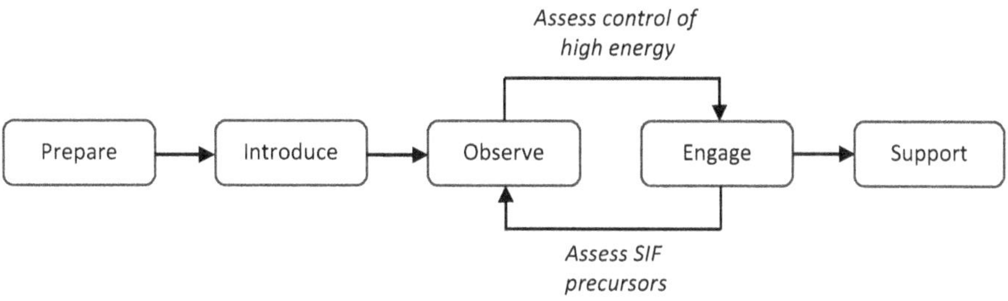

FIGURE 9.1 Key process of a complete SIF-Focused Safety Walk. This process outlines the structure of an effective safety walk, emphasizing the need to cycle between observing work and engaging with workers to assess high-energy hazards and SIF precursors.

OBJECTIVES OF A SAFETY WALK

The original intent of a Safety Walk was to ensure that sites are free of exposure to life-threatening hazards and to verify that workers have what they need to do their job safely and efficiently. Unfortunately, this activity has sometimes turned into a check-the-box exercise, losing sight of its true purpose. This chapter aims to refocus Safety Walks on their core objectives: identifying critical hazards, improving controls, and engaging meaningfully with the workforce.[2]

PREPARING FOR THE SAFETY WALK

A SIF-focused Safety Walk should not be an ad hoc activity. The more a leader can prepare, the greater the chance for success. To be adequately prepared, the leader must gather enough information about the site and have sufficient EBS knowledge to be qualified to perform the assessments and provide recommendations.

Preparation begins with gaining an understanding of the work. This includes basic knowledge of the task, required equipment and tools, common high-energy hazards, and expected controls. They should also familiarize themselves with the schedule and identify any known precursors such as unusual work hours, productivity pressures, or unusual configurations.[3] They should consult the crew's direct supervisor to understand the work's context and any ongoing challenges. For example, if they are visiting a pipeline project with very steep slopes, they should be informed about how this terrain impacts the work. This foresight enables more informed, specific, and meaningful conversations with the crew. If a leader does not fully understand the task or the work's context, they should be prepared to ask thoughtful questions and avoid making assumptions.

In addition to understanding the work, a leader must also gain working knowledge of EBS principles and associated tools. This includes using the energy wheel to perform a complete hazard scan; performing energy assessment to identify high-energy hazards; applying the definitions of Direct and Alternative Controls to determine if protection is adequate; and using SIF precursor questions to assess the likelihood of a critical error during work. With this knowledge, EBS can be naturally woven into observations, discussions, assessments, and recommendations in a way that prioritizes SIF prevention.

STARTING WITH PURPOSE

A prepared leader should begin by introducing themselves and explaining why they are visiting. They should emphasize that the Safety Walk is an opportunity to enhance the control of high-energy hazards and for the company to learn how to better support its employees. It is not a trivial fault-finding exercise.

If possible, it is crucial for the leader to make every effort to observe the Pre-Job Safety Brief (PJSB) before touring the site. This will hopefully answer many common questions that arise during a Safety Walk, such as details about the task, work environment, hazards, controls, and emergency response plans. Observing the PJSB also provides insights into SIF precursors, like the plans for change, knowledge of safe work procedures, and deviations from routine. With this information, the leader can ask more informed questions and make the Safety Walk less disruptive to the crew.

OBSERVING ACTIVE WORK

Observing conditions during active work is the backbone of a Safety Walk. When these observations are focused on the control of high-energy hazards, we call this an Energy-Based Observation (EBO). When conducting an EBO, active work should be observed, which helps the leader understand the nuances of the work and avoid speculation.

An EBO begins by identifying all hazards using the energy wheel. Then, the high-energy icons and energy calculations from Chapter 4 are applied to determine which hazards, if any, meet the high-energy threshold. For each high-energy hazard, the leader must assess whether a Direct or Alternative Control is present by strictly applying the definitions introduced earlier. The objective is to ensure an adequate control is in place for every high-energy hazard.

A Safety Walk is not just an opportunity to evaluate work; it is also a chance to improve conditions. Engaging with the crew can reveal ways the organization might strengthen controls by moving up the hierarchy of energy control (see Chapter 5). For example, questions can be asked about whether a hazard could be eliminated through design or whether a Direct Control could replace an existing Alternative Control. These conversations help translate observations into practical, field-informed improvements.

Observation is essential to a Safety Walk, but it's not the only component. Assessing high-energy hazards and controls often involves engaging with employees, asking questions, and testing assumptions to fully understand and evaluate potential exposures. This is where precursor analysis can be applied.

ENGAGING WITH EMPLOYEES TO ASSESS PRECURSORS

A Safety Walk is also an opportunity to learn about the work context and how it affects the employees. Even with Direct Controls in place, the implementation of these controls depends on people, and their performance and wellbeing are directly linked to the physical, social, and psychological environment that surrounds them. Most often, this work context stems from decisions made long before work began, and as a leader, it is important to understand and correct negative influences.

In Chapter 6, we reviewed the 11 key SIF precursors. For each precursor, three sample questions were provided to help a leader determine if the precursor was present or not. These questions were precisely worded for the sake of the research, but they do not necessarily make a good script for an engagement. My colleague Gregg Slintak put it best: those are not the questions you must ask, but the questions you must be able to answer by the end of an engagement. As with any conversation, these questions should be scattered throughout the Safety Walk as they naturally arise. For example, information about the worker's familiarity with the task can be gleaned by simply asking them how long they have been in the trade and how often they perform the job.

When a Safety Walk starts with an observation of the PJSB, the following precursors can be assessed without any additional questions: quality of the PJSB, plan to address change; knowledge of safe work procedures; and attitudes toward safety. This leaves seven precursors to assess through conversation, which relate to the employee's history with the work (i.e., departure from routine, familiarity with the task, and risk normalization), the status of the project or work order (i.e., productivity pressure and long work hours), and how the workers are feeling (i.e., distractions and fatigue).

A good SIF-focused discussion doesn't have to be formal or rigid. With practice, it can flow so naturally that it is part of a casual and comfortable conversation. Figure 9.2 shows how all 11 precursors can be assessed through a few basic questions

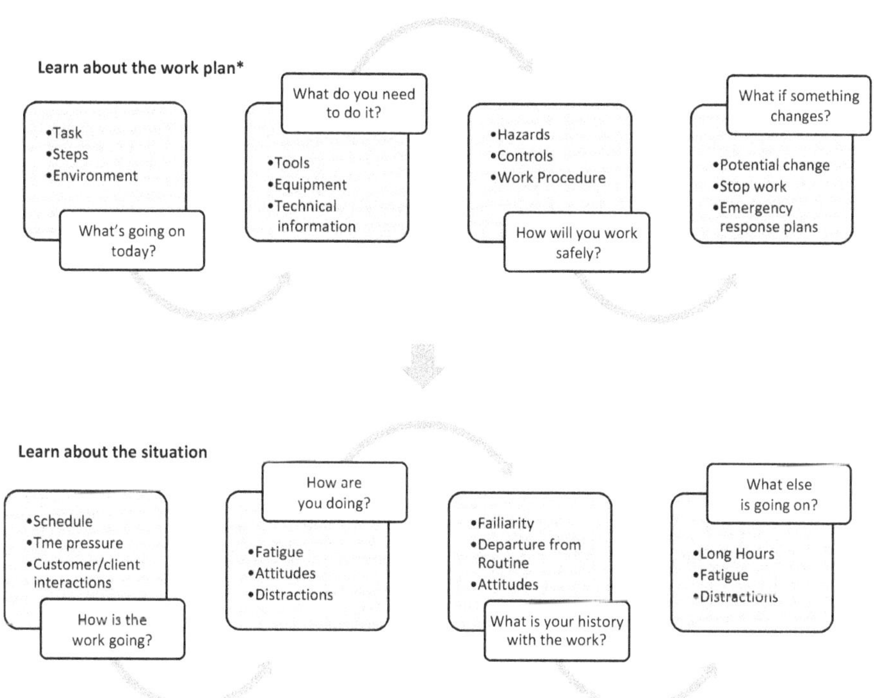

FIGURE 9.2 Conversational approach to assessing SIF precursors during a safety walk. This guide supports natural dialogue that yields the information needed to formally assess SIF precursors. *If a PJSB is observed beforehand, start with the second set of questions focused on the work situation.

that most good leaders already ask. Once again, we see how EBS principles can be woven into the activities we already perform.

Because conditions are never perfect and people bring the realities of their own lives to work, at least one precursor is almost always identified in an engagement. However, since precursors are intended to be anomalies, identifying more than a couple precursors on one job should be rare and alarming. When three or more precursors are present, high-energy work may warrant reassessment or delay due to the significantly increased likelihood of critical error.

Important Connection between Controls and SIF Precursors

The sensitivity to SIF precursors depends on the number of high-energy hazards and the type of controls. If there are only a couple high-energy hazards and each has a Direct Control, the sensitivity to SIF precursors is relatively low. However, if there are many high-energy hazards that are addressed with Alternative Controls that are vulnerable to human error, then the sensitivity to SIF precursors is high. Since the risk of SIF relates to the interplay between hazards, controls, and SIF precursors, the leader must attend to *all three elements* during a Safety Walk.

Guidance for Effective Engagement

As we have integrated precursor analysis into Safety Walks and emphasized meaningful engagement, a few key practices have emerged.

- ***Know the work and the workers***. To get candid responses and complete information, leaders should understand the work and the workers. At a minimum, they should establish rapport before asking questions. Crews may feel anxious due to past compliance audits so leaders should clarify they are there to learn and support, not judge. Reducing this angst will encourage open and honest conversations.
- **Look for and celebrate the positives**. On the surface, engagements can seem like an interrogation. Reframing it as a process of identifying strengths can be more effective. Since there are usually more positive attributes than problems, strengths should be recognized and commended while addressing weaknesses. This shift in thinking improves how leaders ask questions and how workers respond.
- **Recognize that most problems stem from deficiencies in the system, not personal flaws**. When employees feel safe opening up, it provides an opportunity to learn how work is actually performed. Although it is tempting to find fault with employees when a precursor is identified, most issues stem from upstream decisions, resource limitations, impractical work procedures, and human nature. For example, we may discover uncontrolled exposure to adjacent traffic because a physical barrier is too expensive, crews deviating from standard procedures because the steps are impractical, and personal factors stemming from everyday life. These issues are not the

fault of the workers but still must be understood and addressed because they are drivers of SIFs.

- **Look for anomalies**. Precursors are anomalies that differentiate normal work from work with elevated SIF potential. As we found in the research, we have a tendency to overinflate weak signals, sounding alarm bells for every minor issue. However, there are weak signals present in almost all normal work. For example, there is almost always a little schedule pressure, fatigue, or distraction. Instead of reacting to weak signals, leaders should focus on strong signals such as extreme schedule pressure from client demands or outages, very long hours, and deviations from routine that involve new high-energy hazards.
- **Be sensitive to personal issues and the desire for privacy**. It may be appropriate to have some conversations privately. For example, I remember one conversation where an employee was asked if anything outside work might cause them to be fatigued or distracted. The employee's response was, "No, I leave my family problems at home." Anyone with the slightest empathy can hear that the employee has a family problem. If the employee has issues that could compromise their ability to safely control high-energy, their supervisor should know and work with them privately to provide support and create an alternative plan.
- **Act when necessary**. Sometimes a Safety Walk will reveal a serious condition that requires immediate action. For example, there could be an uncontrolled exposure to a high-energy hazard and multiple precursors present in one job. In these extreme cases, work should be halted until the issues are corrected.
- **Provide recommendations and support**. In the spirit of continuous improvement, ideas and constructive criticism should be provided directly to workers and field supervisors. When providing this feedback, leaders should be respectful but authoritative, emphasizing the urgency of controlling high energy. They should also rephrase what they learned and summarize how they will address employee concerns.
- **End by asking, "How can I help?"** When there is clear personal commitment to improving things, it will encourage the crew to share ideas and demonstrate that the leader is a partner.

CASE STUDY OF A SIF-FOCUSED SAFETY WALK

A case study is provided to demonstrate the analysis portion of a typical Safety Walk. This case study illustrates how hazard review, control verification, and precursor analysis can be integrated during a Safety Walk to produce a holistic, SIF-focused assessment. As shown in Figure 9.3, the case involves site preparation with simultaneous excavation and survey operations. The picture shows routine work that has both safe and risky elements. Since a picture alone conveys limited information, we will also include insights gathered from discussions with the crew and observations made during the work.

Based on all the information gathered, we performed the analysis shown in Table 9.1. This table includes the identified high-energy hazards, the presence or absence of Direct Controls, and a brief description of each control.

FIGURE 9.3 Workers perform trenching and surveying activities during site preparation. Six high-energy hazards are labeled corresponding to Table 9.1.

Assessing the Control of High-Energy Hazards (Energy-Based Observation)

The energy-based observation (EBO) had two steps: (1) identifying high-energy hazards and (2) assessing the controls. In the case study image, there are several obvious high-energy hazards. For example, there was an excavator and a deep trench.

TABLE 9.1
High-Energy Hazards and Direct Controls Assessment for the Case Study in Figure 9.3

Number	High-Energy Hazard	Direct Control Description	Direct Control?
1	Excavator	None	No
2	Dump truck	None	No
3	Work in 7-ft deep trench	Engineered trench box	Yes
4	Underground electrical utility line (electrical contact)	De-energization, zero-voltage test, and grounding	Yes
5	Underground electrical utility line (arc flash)	De-energization, zero-voltage test, and grounding	Yes
6	Work near 7-ft high leading edge	None	No

However, it wasn't until the leader engaged with the crew that the following critical details were clarified:

- There are underground electrical utility supply lines that were properly located, de-energized, tested for zero voltage, and grounded;
- A trench box was installed before work was being performed in the trench;
- Excavations were ongoing and dump trucks were regularly entering and exiting the site; and
- There was no thunderstorm in the forecast.

Based on the information gathered, the analysis revealed six high-energy hazards, three of which had a Direct Control and three did not. The three that did not have a Direct Control also did not have sufficient Alternative Controls. For example, there was no protection around the trench to prevent falls and no spotter to monitor adjacent activities.

With this information, the leader commended the crew for executing the proper de-energization procedure for the electrical supply and for using trench boxes anytime workers entered the excavation. Then, attention turned to the uncontrolled high-energy hazards. For the potential for fall into the trench (hazard 6), the team determined that a fall protection system was not feasible for this task and site configuration. Similarly, for the heavy mobile equipment (hazards 1 and 2), they learned that the *company* does not have a Direct Control or an established procedure for working on foot alongside heavy mobile equipment. It was also deemed too costly and cumbersome to install concrete barriers for a job of this duration. Therefore, for these three uncontrolled high-energy hazards, the leader brainstormed Alternative Controls with the crew.

To control the fall into the trench, cones with cone toppers were used to create a perimeter around the unprotected edges and a sign was installed to alert workers of the fall hazard. For the heavy mobile equipment, a pedestrian pathway was marked with caution tape and one of the experienced crew members served as a spotter to ensure that workers were in this exclusion zone when the equipment was operating. Finally, the leader escalated concerns to the company, noting the presence of multiple life-threatening exposures in a common job and the lack of feasible Direct Controls.

In Chapter 10, we will explore how the hazard and control assessment in an EBO can yield a new form of safety measurement.

Learning about Work Context with SIF Precursors

During the conversations with the crew, the leader learned a few things about the work that were not apparent from the EBO:

- The PJSB discussion was complete. The team used the energy wheel for hazard recognition, there was specific acknowledgement of high-energy hazards, they discussed a plan for change, and everyone participated.
- The crew members are knowledgeable about the work but not overconfident.

- The crew was two days behind schedule already and was "playing catch-up" to be ready for the concrete pour that was planned for later in the week.
- The employee working adjacent to the excavation had a sick child at home, and that was weighing on his mind.
- The crew members were slightly distracted by the site traffic, but it was not extreme.

From this information, the leader determined that there were two SIF precursors present: distraction and schedule pressure. These were particularly concerning because there were multiple high-energy hazards without a Direct Control. To address these precursors, the leader reassured the crew that safety comes before productivity and that management will handle any consequences of a delay. The leader also worked with the supervisor to find a different job for the distracted laborer, so he was not directly exposed to both the leading edge and the heavy mobile equipment.

This case study doesn't represent a perfect Safety Walk nor a perfect response. However, it does show how a Safety Walk can be SIF focused and how a leader can take both short- and long-term actions to improve conditions. When the leader ended the visit by explaining what they learned and that they had already made recommendations to the company, the crew felt that their voices were heard, and their insights mattered.

BEING AN EFFECTIVE ONSITE SAFETY LEADER

How you do a Safety Walk is just as important as *what* you do.

In many cases, Safety Walks are conducted by leaders who are deeply familiar with the work, like a site manager or foreman who walks the job daily. These individuals often know the crew and understand the work. Their presence is expected, their language is familiar, and their leadership is embedded in day-to-day operations.

But sometimes, a Safety Walk is conducted by someone more removed from the field such as an executive, senior manager, or safety professional. In these moments, the dynamic shifts. Workers take note, and scrutiny is higher. How the leader behaves often sends a clear signal about the company's values and priorities. What might seem like a casual comment can leave a lasting impression. The leader's tact, tone, and presence matter as much as the content of the Safety Walk itself. If EBS is the seed of SIF prevention, and safety systems are the soil, then intentional, visible leadership is the sunlight that brings the system to life.

Recognizing that a Safety Walk led by someone outside the day-to-day operations carries a different kind of weight, we set out to define what makes these engagements truly effective. While many leadership principles apply broadly, the following guidance is tailored specifically for those who don't work on-site every day:

- **Look the part**. Wear the appropriate jobsite attire and follow site rules to emphasize camaraderie. For example, consider how wearing an expensive suit and new shiny boots would be perceived compared to standard coveralls and worn work boots.

- **Be attentive**. Be physically and mentally present and put your phone or other distractions away during the engagement. For example, consider how a worker would feel if you were checking your phone while they are sharing their ideas.
- **Ask questions**. Show humility and recognize that employees are the experts in their work. Be inquisitive and give the employees the opportunity to showcase their knowledge and give them time to explain without rushing to the next question.
- **Use active listening skills**. Ask follow-up questions, repeat key feedback, and give visual cues such as head nodding and maintaining eye contact.
- **Ditch the clipboard**. Nothing sends the message of authority and superiority more than walking around with a clipboard. Demonstrate the desire to learn rather than judge by recording the results after the engagement is complete.
- **Involve the crew supervisors**. As the link between company leadership and front-line employees, crew supervisors are the glue that holds SIF efforts together. When engaging with the crew, involve the supervisor as a partner to avoid miscommunication, strengthen alignment, and reinforce their authority.

By demonstrating these key attributes, leaders can show that they not only care about SIF prevention, but they also care about their employees.

CONCLUSION

Safety Walks serve many different purposes, from control verification to culture building. Too often, leaders reduce this activity to a fault-finding and a check-the-box exercise, which is a lost opportunity and a shameful waste of resources. At its essence, a Safety Walk should be seen as an opportunity for leaders to check and learn, and they are an ideal activity for EBS integration.

In design and planning, we focused on preventing SIFs by anticipating and controlling high-energy hazards before work begins. Safety Walks help verify whether plans are being executed effectively and identify gaps requiring company action. It is well known that work as planned is not always the same as work as performed. Safety Walks provide a way to understand these gaps and how to improve the control of high-energy hazards.

Integrating EBS principles into Safety Walks enables leaders to hone attention to a small number of very important hazards and conditions. High-energy focuses on the few hazards with serious potential, the controls definitions offer a formal way to assess if work is safe, and the precursor questions help to ensure that the work context is conducive to human performance. When leaders wrap this together by demonstrating genuine care for their employees and their ideas, the culture of SIF prevention is strengthened.

In the next chapter, we will explore how Safety Walks can yield critical information that can be used for organizational learning and to transform safety measurement from measuring and responding to lagging indicators to innovative methods of monitoring and control.

KEY TAKEAWAYS

1. Safety Walks bridge the gap between planning and execution, offering an opportunity to learn from and improve normal work.
2. EBS integration ensures Safety Walks focus on SIF prevention.
3. A holistic Safety Walk involves observing and engaging with a crew to assess high-energy hazards, controls, and precursors.
4. Safety Walks give leaders a chance to demonstrate company commitment to SIF prevention.
5. Collaboration with crew supervisors strengthens alignment and supports effective field operations.

NOTES

1 Salas, R., & Hallowell, M. (2016). Predictive validity of safety leading indicators: Empirical assessment in the oil and gas sector. *Journal of Construction Engineering and Management*, 142(10), 04016052.

2 Kines, P., Andersen, L. P., Spangenberg, S., Mikkelsen, K. L., Dyreborg, J., & Zohar, D. (2010). Improving construction site safety through leader-based verbal safety communication. *Journal of Safety Research*, 41(5), 399–406.

3 Alexander, D., Hallowell, M., & Gambatese, J. (2017). Precursors of construction fatalities. II: Predictive modeling and empirical validation. *Journal of Construction Engineering and Management*, 143(7), 04017024.

10 Measuring and Monitoring Safety with High-Energy Control Assessments (HECA™)

Measurement is the cornerstone of both knowledge and innovation, transforming abstract ideas into quantifiable insights. In the realm of safety, and especially in energy-based safety (EBS), measurement allows us to quantify the forces and controls that define SIF risk, giving us a tangible way to understand and manage what would otherwise be intangible. Metrics provide the foundation for benchmarking with peers, testing the effectiveness of interventions, and making informed decisions. Without them, we would be navigating the complexities of safety without a compass, unable to set objective goals, measure progress, or work cohesively as a team. As James Vincent eloquently said, "measurement can be a tool for social cohesion and control, letting us coordinate individual efforts into something greater than the sum of its parts." However, when misused, metrics can also divide, turning collaboration into competition and shifting the focus from learning to manipulation.

The safety metrics we choose directly reflect our values and priorities. When we concentrate on a set of numbers, we are not merely counting numbers; we are defining what is important and, by extension, what is not. Therefore, leaders must be intentional and report metrics that advance toward the organization's most important goals.

Equally crucial is how we use the safety metrics we choose. Metrics generally serve two purposes: comparison and learning. Ideally, they should support both, but in practice, the focus has been heavily skewed toward comparison, asking questions like: "*How did we perform last year compared to our peers?*", "*Did we improve as a company since last quarter?*" or "*Which contractor has better safety performance?*" Most companies use historical injury rates to answer these questions, driving important business decisions such as resource allocation, promotion, and incentives. Unfortunately, as we will unpack in this chapter, injury rates are abysmal indicators of safety performance because they contradict modern philosophy, lack statistical validity, and drive reactive decision-making.

This chapter begins by examining the deep-rooted flaws in traditional safety metrics that have stalled progress in SIF prevention. Although moving away from outdated metrics is long overdue, finding viable alternatives has proven to be a persistent and stubborn challenge. Enter High-Energy Control Assessments (HECA™), the crown jewel of EBS. HECA represents a paradigm shift, offering a structured and scientifically grounded approach to measuring safety as the presence of safeguards.

DOI: 10.1201/9781003641483-12

Designed to assess the extent to which high energy is effectively controlled during active work, HECA bridges the gap between lagging and leading indicators, providing an actionable method to monitor both success and exposure. With HECA, organizations can move beyond outdated metrics, focusing instead on what truly matters: controlling high-energy to reduce SIF risks. This transformation in safety measurement begins by revisiting the very definition of the word "safety."

WHAT IS SAFETY?

For decades, safety was narrowly viewed as the absence of injuries, with a work period considered safe if no one was hurt and unsafe if an injury occurred. This view of safety is reflected in injury rates, zero-injury goals, and signs reporting time since the last injury. A hyperfocus on these outcomes has led to an environment where injuries are hidden, and learning is stifled.[1]

There has been a refreshing shift in the way safety is defined, captured by Todd Conklin's statement: "*Safety is not the absence of events; safety is the presence of defenses.*[2]" This idea is now widespread, appearing in email signatures, posters, and executive communications, and even embedded as a core value in some company websites and ESG disclosures. Yet, as a recent study revealed, many of these same companies continue to report injury rates and reward employees for meeting injury-based targets.[3] Despite embracing new principles, metrics still reflect the outdated view they claim to have moved beyond. If we say that safety is the presence of safeguards, isn't that what we should be measuring? HECA enables us to do just that.

HOW SAFE IS THIS JOB?

Below is a simple activity to help leaders reflect more deeply about safety metrics. Considering the image of normal, everyday work depicted in Figure 10.1: *How safe*

FIGURE 10.1 Photo of normal work.

is this job? Take a moment and give it some thought. Can you describe the extent to which it is safe or unsafe? If you had to put a number on it, what would it be?

Responses to this question are typically vague and hesitant with phrases like "not very safe" or "could be better." Clearly, the job is neither entirely safe nor entirely unsafe. Like virtually all work, it is somewhere in between, making it challenging to assign a number.

This small activity illustrates an important point: we do not have a method of measuring safety in the present tense.

If we were to use the prevailing method of safety measurement (injury rates), we would need to wait until the job is complete, count how many injuries occurred, divide by the number of worker-hours expended, and normalize by a factor like 200,000. For small jobs, where the likelihood of injury is negligible, the injury rate would be zero, falsely implying perfection. Since we know the work isn't perfectly safe, the injury rate misrepresents reality and creates a logical fallacy. We need a metric that measures safety in real-time, reflecting the extent to which safeguards are actively in place.

Before we can move forward with a new metric like HECA, we must first challenge the dominance of traditional injury metrics and put them in their proper place.

THE TYRANNY OF TRIR

No metric is as pervasive or influential as Total Recordable Incident Rate (TRIR).[4] Used as the primary measure of safety performance for nearly 50 years, TRIR measures the frequency with which a company experiences a recordable incident, scaled per 200,000 worker-hours (see Equation 10.1). In the US, it is based on a standardized definition of "recordable" incidents, established under the Occupational Safety and Health Act of 1970. According to OSHA, an incident is recordable if it results in a work-related injury or illness requiring medical treatment beyond first aid leading to loss of consciousness, days away from work, restricted work, or job transfer.

$$\text{TRIR} = \left(\text{Number of Recordable Incidents} \times 200{,}000\right) / \left(\text{Number of Worker - Hours}\right) \qquad (10.1)$$

For example, a company that accounts for 3 recordable incidents over 350,000 worker-hours in a month would have a TRIR of 1.71 per 200,000 worker-hours for that month.

TRIR is used for many business purposes, such as reporting outcomes to the Board, benchmarking against peers, prequalifying contractors, assessing managerial performance, and gauging the effectiveness of safety programs. It is especially important to executives because it can affect workers' compensation premiums and reputation and is closely examined by potential clients or investors. While other lagging indicators are also used, none are as widespread or impactful as TRIR and its variants. Therefore, any discussion about alternative safety metrics must begin with a thorough indictment of this obsession.

Structural Problems with TRIR

If TRIR were even half as valid as it is popular, safety measurement wouldn't be such a persistent challenge. While TRIR is widely adopted, it suffers from several fundamental flaws that undermine its validity. Laying out these flaws one after another showcases its absurdity and reveals an undeniable need for progress. TRIR will be used as the example here because it is the primary offender; however, these structural problems extend to most other injury rates, such as lost-time (LTI) rates, days away from work or job transfer (DART) rates, and even fatality rates.

- ***TRIR is antithetical to the modern definition of safety as the presence of safeguards***. Since TRIR is a binary metric, with 1 representing injury and 0 representing all other work, TRIR is designed to measure safety as the absence of injuries. The metric provides no information about whether safeguards were in place or if an outcome was a result of luck or misfortune.
- ***TRIR is skewed by underreporting***. Since organizations are incentivized to keep injury rates low to appear safer, there is often considerable pressure to underreport injuries. Research shows that underreporting of injuries is rampant, with some studies suggesting that over half of all recordable injuries are never reported.[5] Further evidence of underreporting has emerged as companies remove TRIR from their compensable metrics. As employees seek treatment for incidents that they previously concealed, TRIR skyrockets even though the company is safer than ever. Therefore, companies with compensable TRIR targets simply do not know how many recordable injuries occur in their business.
- ***TRIR does not account for the severity of incidents and diminishes the importance of SIFs***. A minor injury such as a two-stitch cut is counted the same as a fatality even though the impact of these events is drastically different. This binary approach, which treats all recordable injuries equally, overlooks the relative severity of incidents and can lead to an obsession with low-consequence events. This misplaced focus takes attention away from SIFs, which are comparatively rare but infinitely more impactful.
- ***TRIR promotes reactive behavior***. Since TRIR only counts incidents *after* they occur, it encourages reactive behavior. A sudden increase in injury rates often drives companies to intensify general safety efforts such as observations or audits beyond what the company can sustain. As the company relaxes when injury rates are favorable, injury rates rise again, and the cycle continues. Unfortunately, we find that lagging indicators drive leading indicators, rather than the other way around.[6]

We have known of these structural problems of injury rates for years. However, with rose-colored glasses, some clients, board members, and company leaders make the case that injury rates have "worked well in the past" and that the injury rates "are what they are." However, armed with hard data, we are now able to make the argument that using injury rates for business decisions is both structurally flawed and statistically invalid.

Statistical Problems with TRIR

Critics of injury rates have long argued that injuries occur so infrequently that they lack statistical stability and reliability. This issue becomes particularly evident in short timeframes, such as when a team of two completes a 4-hour task (8 worker-hours), where calculating a TRIR would clearly be meaningless. But how many worker-hours are required for TRIR to be statistically meaningful? We deployed a team of clients, contractors, and researchers to answer this question using hard empirical data.

With the size and strength of the CSRA community behind us, we collected monthly recordable injury data spanning 15 years, resulting in a dataset of 3.2 trillion worker-hours. We examined the data for dependence (e.g., is a recordable injury more or less likely after a previous incident and, if so for how long?) and the extent to which the timing of a recordable injury is random (e.g., is there some pattern to the timing of a recordable incident or do they occur randomly over time?).[7]

This examination revealed three key results:

1. ***TRIR of the past does not predict TRIR in the future***. Companies with a relatively low TRIR in the past do not necessarily have a low TRIR in the future, and vice versa.
2. ***TRIR has no statistical association with SIFs***. Trends in recordable injuries have no bearing on a company's risk of a SIF because recordable injuries and SIFs follow different patterns and occur for different reasons. For example, preventing twisted knees and small cuts to the finger has no logical or statistical connection with preventing explosions or dropped loads. This disconnect explains why many companies experience perplexing SIFs despite a very low TRIR. Statistically, TRIR is not a proxy for SIFs, which further debunks the assumptions associated with the infamous injury pyramid.
3. ***TRIR is about 98% random, so it is statistically meaningless unless it is derived from over 10 million worker-hours***. The timing of a recordable injury is almost perfectly random and there are no feedback loops or predictable patterns. This explains why companies with stable safety systems often experience long stretches without recordable injuries followed by sudden, inexplicable clusters. SIFs, which are hundreds of times rarer than recordable injuries, are equally random. Therefore, to achieve statistical stability, SIF rates would require data from hundreds of millions or even billions of worker-hours. This makes SIF rates an unsuitable alternative to TRIR. The Dice Analogy provided below helps communicate the effects of probabilities and randomness on outcomes.

THE DICE ANALOGY: WHAT DOES IT MEAN TO BE RANDOM?

To demonstrate the effects of randomness, I ask leaders to roll a 6-sided die 10 times and provide the total sum of their rolls. While some may have a high number like 48, others may have a lower number like 27. Although it may seem like it from the numbers alone, no person is a better roller than another.

> Because of random effects, we can get vastly different results even if the system (6-sided die) is the same. We simply haven't rolled the dice enough times to yield similar results.
>
> The same is true for TRIR. Even if we were to do the exact same work with the exact same safety processes, we can observe vastly different results from one month, quarter, or even year to the next. Only, unlike a 6-sided die, safety is like a 200,000-sided die with 199,999 zeros and only one 1. The research indicates that, from a statistical perspective, we roll this die once every worker-hour and, to understand what the system is designed to produce, we must roll the die over 10 million times.

The research provides bold evidence that businesses need to move away from making consequential business decisions from such a volatile metric. They must critically rethink how TRIR is applied and eliminate practices that perpetuate its misuse. To move forward, leaders should consider the following:

1. *Stop overreacting to short-term variations in TRIR.* Because TRIR is rare and random, short-term fluctuations should be expected. For almost all companies, a one-month spike in recordable injuries doesn't mean that the safety system is suddenly flawed, it is merely bad luck following a streak of good luck. Companies must accept volatility and resist overreacting to noise.
2. *Stop rewarding and penalizing random variation.* Since the average company requires tens of millions of worker-hours to report a TRIR to one decimal point, they should be very careful making *any* comparisons using TRIR. For example, most companies don't have enough worker-hours to detect statistically significant changes from one year to the next, let alone one month to the next. Consequently, TRIR should not be used to make decisions related to bonuses and other incentives, or for contractor prequalification. When we do this, we are only rewarding or penalizing random variation, not some meaningful change to the system.
3. *Stop using TRIR as a proxy for SIF.* For years, many assumed that fatalities were being addressed by virtue of the declining trends in TRIR. However, as the data clearly show, this idea holds no scientific merit. SIFs must be addressed separately through targeted interventions that help to control high-energy hazards.

TRIR Isn't All Bad

Before transitioning to alternatives, we must acknowledge that TRIR has endured for nearly a century for an important reason: it is the one thing in safety that we do the same way. TRIR and its variants are typically based on a single, common, and standardized definition of a "recordable" injury that is governed by a regulator. This level of consistency enables a unique apples-to-apples comparison across businesses, sectors, and even industries. Other alternative measures such as leading indicators are not standardized across companies, making them useless as a comparable metric. To be fully successful in counterbalancing TRIR, alternative metrics must be consistently applied.

LEADING INDICATORS TO THE RESCUE?

After we completed the TRIR study, I had frequent opportunities to present the findings to company leaders. They mostly listened attentively and asked thoughtful questions, which gave me the impression of genuine interest in understanding the implications. Inevitably, however, someone would pose a challenging but fair question: "If not TRIR, then what?" Within the safety profession, the common answer has been "leading indicators!" While promising, leading indicators have significant limitations that prevent them from being a wholesale solution.

Leading indicators are satisfying because they are somewhat predictive and measure activities that are under our control like safety observations or leadership engagements. Companies can create their own set of leading indicators by focusing on the activities that are most important to them, devising methods to measure these activities, setting goals, and measuring progress toward the goals. Leading indicators are an excellent way to look *within* a business.

Unfortunately, leading indicators are not yet viable for looking *across* businesses because they are not standardized, and perhaps never will be. While safety professionals often take pride in their leading indicators, these metrics often have little weight at higher levels of the organization because Board members and executives must also assess the company's position relative to peers (it's not just their preference, it's their job). Without consistency in how companies conduct safety activities and report associated metrics, external comparisons are invalid.

While leading a team of electric utility companies to standardize a small set of leading indicators, we encountered a significant and unexpected challenge causing the initiative to fail. Despite operating in a highly regulated and collaborative sector, there were substantial differences in how each company defined, executed, and measured safety activities, with these practices deeply embedded in their individual cultures. It became evident that achieving alignment to one standard approach was not feasible for *existing* practices. This redirected our focus toward developing something entirely new.

HIGH-ENERGY CONTROL ASSESSMENTS (HECA)

Recognizing the limitations of existing metrics and the challenges of standardization, we turned our attention to creating a fundamentally new approach to safety measurement. HECA emerged as a solution, designed to focus on measuring and monitoring the control of high-energy hazards.[8] By grounding HECA in EBS principles and incorporating clear, objective rules for assessment, we sought to overcome the shortcomings of traditional metrics. Specifically, our goal was to create a metric that is:

- Aligned with the definition of safety as the presence of safeguards
- Focused on serious injury and fatality (SIF) conditions
- Statistically stable because it can be assessed frequently
- Based on a set of well-defined rules
- Underpinned by the science of energy-based safety

Baking all these features into one assessment strategy was a bold and challenging undertaking, but one that proved to be effective. The result of our efforts was a new approach to safety monitoring that we called HECA. It is a new way of assessing safety performance as the extent to which front-line workers are protected against life-threatening hazards. HECA is specifically calculated as the percentage of high-energy hazards with a corresponding Direct Control (see Figure 10.2 and Equation 10.1). HECA is binary, categorizing each condition as either success (a high-energy hazard with a corresponding Direct Control) or exposure (a high-energy hazard without a Direct Control).

$$\text{High} - \text{Energy Control Assessments}\left(\text{HECA}\right)\text{computation}$$
$$\text{HECA} = \frac{\text{Success}}{\left(\text{Success} + \text{Exposure}\right)} \tag{10.1}$$

***Success*:** total number of high-energy hazards observed with a corresponding Direct Control
***Exposure*:** total number of high-energy hazards observed without a corresponding Direct Control

Like TRIR, HECA is based on hard rules that involve limited interpretation. Determining whether a hazard is high-energy or not is based on physics, and the adequacy of the controls is based on the Direct Control rules. If the rigor of energy and control analysis is preserved, HECA enables uniform assessment both within and across organizations.

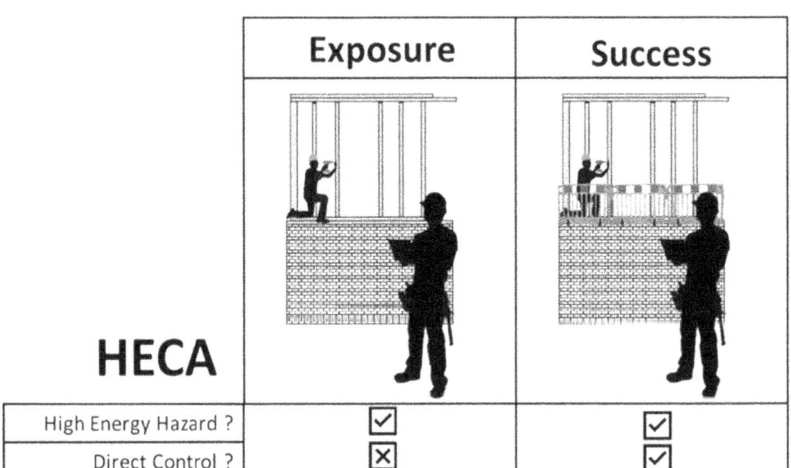

FIGURE 10.2 HECA metric structure. This figure illustrates the visual distinction between success and exposure in high-energy work.

This figure appeared in: Erkal, E. D., & Hallowell, M. (2023). Moving beyond TRIR: Measuring & monitoring safety performance with high-energy control assessments. *Professional Safety*, 68(05), 26–35. Reprinted with permission.

You may notice that Alternative Controls are not included in HECA. This is by design. Since our ultimate aspiration is to have a Direct Control for every high-energy hazard, Alternative Controls are not currently counted in the HECA metric. However, leaders should always ensure that an Alternative Control is present if a Direct Control is not feasible, regardless of the impact on the HECA score. As HECA evolves, consideration will be given to Alternative Controls and how they may be incorporated into future versions of the metric.

HOW TO IMPLEMENT HECA

HECA was designed to be a product of the Energy-Based Observation (EBO). As discussed in Chapter 9, an EBO is the part of a Safety Walk that is focused on the identification and assessment of high-energy hazards and their corresponding controls. If done well, an EBO yields the necessary information needed to compute a HECA score.

Equipped with the information from an energy-based safety walk, HECA can be completed through three steps:

Step 1: Identify high-energy hazards, i.e., those with enough energy that a serious injury or fatality is the most likely outcome if contacted (i.e., 1500 Joules or more). The high-energy icons are often used for efficiency.

Step 2: Assess whether each high-energy hazard identified has a corresponding Direct Control at the time of assessment or not. High-energy hazards with a Direct Control are marked as "success" and those without a Direct Control are marked as "exposure."

Step 3: Compute the HECA score by dividing the total number of high-energy hazards with a Direct Control (success) by the total number of high-energy hazards observed (success plus exposure).

CASE EXAMPLE OF HECA

Let's revisit Figure 10.1 (the photo of normal work) and the case study from Chapter 9 in the context of HECA. As a reminder, the safety walk yielded the high-energy hazards and control assessments summarized in Figure 10.3. This assessment was based on the following information collected through an engagement with the employees:

- There are underground electrical utility lines that were properly located, de-energized, tested for zero voltage, and grounded;
- A trench box was installed when the work was being performed in the trench;
- Excavations were ongoing and dump trucks were regularly entering and exiting the site (so they were considered site traffic, which requires different controls from the excavator); and
- There was no thunderstorm in the forecast.

Number	High-Energy Hazard	Direct Control Description	Direct Control?
1	Excavator	None	No
2	Dump truck	None	No
3	Work in 7-ft deep trench	Engineered trench box	Yes
4	Underground electrical utility line (electrical contact)	De-energization, zero-voltage test, and grounding	Yes
5	Underground electrical utility line (arc flash)	De-energization, zero-voltage test, and grounding	Yes
6	Work near 7 ft-high leading edge	None	No

FIGURE 10.3 HECA case example. This example presents the results of an energy-based observation (EBO) performed in a Safety Walk, showing identified high-energy hazards and whether each has a corresponding Direct Control in place.

The HECA score for this case study would be computed as follows:

Step 1: *Identify High-Energy Hazards*
A total of six high energy hazards were identified. When assessing if a hazard is high-energy, the high-energy icons were used first. If the hazard was not represented by an icon, a formal energy computation was performed to determine if the hazard was high-energy (1500 Joules or more).

Step 2: *Assess Direct Controls*
An assessment was then performed for each high-energy hazard to determine if there was a corresponding Direct Control. For future data analysis and intelligence, the relevant controls (both present and absent) were recorded. The assessment revealed that three high-energy hazards had corresponding Direct Controls (success) and three did not (exposure).

Step 3: *Compute HECA Score*
The HECA score is then computed by dividing the number of high-energy hazards with a Direct Control by the total number of high-energy hazards observed. In this case, the HECA score is 3/6 or 50%.

This case example illustrates a harsh reality: our work is far from perfectly safe. Each one of these exposures is more than just a number; it is a condition where one mistake can cost someone their life.

WHY HECA SCORES MIGHT BE TOUGH FOR LEADERS TO SWALLOW?

HECA may pose a challenge for leaders accustomed to viewing safety in absolutes. Many companies have established norms where zero injuries or 100% compliance are the only acceptable outcomes. However, our research consistently shows that the average HECA scores among mature companies in North America is between 60 and 70%.[9] This indicates that roughly one-third of life-threatening hazards lack adequate control, often because existing controls fail to mitigate energy effectively or remain susceptible to human error. In some cases, the industry lacks feasible Direct Controls for certain high-energy hazards (e.g., heavy mobile equipment operating near workers on foot often presents an insurmountable challenge). Considerable effort is still required to develop Direct Controls for these complex scenarios.

A HECA score between 60 and 70% will not look good on traditional corporate scorecards. Although HECA reflects reality, it will be uncomfortable initially. The transition to measuring safety as the presence of safeguards will likely be bumpy, making learning and benchmarking across collaborating companies more crucial than ever. By pinpointing SIF conditions that exist in our daily work, HECA has helped us understand why SIFs have been such a widespread issue.

HECA RECORDKEEPING

Robust HECA recordkeeping is critical to extracting useful intelligence. Although it may be efficient to record HECA as one number (i.e., % high-energy hazards with corresponding Direct Controls), recording the specific observations for each high-energy hazard enables meaningful trending. Therefore, we have suggested the following fields to make the most of initial HECA recordkeeping:

- List of high-energy hazards observed (e.g., suspended load, work over 4 ft of height, computation of energy magnitude).
- For each high-energy hazard, was a corresponding Direct Control observed (i.e., yes or no for each high-energy hazard)?
- For each hazard *with* a Direct Control, which control was observed (e.g., engineered rigging, fall protection)?
- When a high-energy hazard *did not* have a Direct Control, what control was missing?

For most companies, collecting these data would not be a large departure from current processes. After all, Safety Walks are a core activity and recording hazard observations is a staple of data collection. Since HECA includes formal assessment and strict rules, training and technological alignment are the most resource-intensive elements of implementation. In return, HECA data can provide valuable insights into SIF-related conditions, enabling proactive intervention before a serious incident occurs.

Many leaders will wonder, "how much HECA data do I need to have a stable metric?" The answer to this question lies in representative statistical sampling, which depends on the number of working crews in the company and the desired degree of confidence. This sampling guidance, along with user-friendly tables, is provided in the Edison Electric Institute (EEI) HECA report.[10]

EXAMPLE INTELLIGENCE FROM HECA

When it comes to learning, a safety metric is only as useful as the intelligence it provides. Since HECA is based on the assessment of ongoing work, it supports both tactical response and long-term strategic planning. For instance, detecting an uncontrolled high-energy hazard during a safety walk can trigger immediate on-site coaching and problem-solving, leading to real-time safety improvements. Over time, tracking HECA as a performance metric offers valuable insights into the effectiveness of high-energy controls, while trends in hazards and controls promote organizational learning, inform research and development, and guide resource allocation.

Since HECA is integrated into routine safety observations and can be analyzed for each individual hazard, relatively large datasets can be generated over a short time. This statistical stability marks a significant departure from relying on rare and random injury counts. When organizations collect meta-data, such as specific high-energy hazards and the presence or absence of corresponding controls, HECA can be unpacked to answer important questions and identify needs for targeted safety investments. These questions may include: Which high-energy hazards are relatively well controlled, and which are not? Which controls are most commonly missing? How do different projects or business units compare in controlling high-energy hazards? How closely do targeted interventions correlate with improvement in hazard control? And, to what extent do HECA trends predict future performance? Examples of this type of analysis are available in the EEI HECA report.

EFFORTS TO PROTECT HECA AS A STANDARD

Companies frequently modify new safety concepts to create a sense of ownership and fit them into their unique cultures. While this adaptation is important for acceptance and integration, it erodes our ability to compare data, aggregate insights, and foster collective learning. Customization is particularly problematic for metrics where comparison and benchmarking are required.

The widespread adoption of TRIR highlights the power of standardization. Its consistent use allows for reliable comparison and shared understanding. HECA must replicate this strength by sticking to a unified definition and set of rules.

If organizations start tailoring HECA to suit their preferences, its value for shared learning and its role in eliminating SIFs, will be significantly reduced.

To promote standardization of HECA, we have taken three key actions:

1. Created and maintained the HECA rulebook
2. Created a HECA governance body
3. Trademarked the term HECA[11]

Rulebook

Even with strict definitions of high-energy hazards and Direct Controls, variability in assumptions, boundaries, and processes could lead to vast inconsistency in application and the inability to compare HECA across companies. To promote consistency, ten HECA rules were established that enable calibration and consistency within and among companies.

HECA RULES

1. HECA is based on observations of active work and not the assessment or classification of an incident or the review of paperwork alone.
2. Each HECA measurement must correspond to one crew performing one task during one working day. Multiple crews and/or tasks require multiple HECA measurements.
3. If a crew performs more than one task during the observation period, a separate HECA measurement must be made for each task.
4. When two or more crews are working in proximity to one another and performing the same task, they must be grouped as one HECA measurement.
5. If two or more hazards have the same high-energy source and the same Direct Control, they must be combined as one entry.
6. If there is a deficiency or missing coverage of a Direct Control, the entry must be recorded as Exposure.
7. Assessments must only be made based on work as it is observed. Hypothetical, anticipated, or speculated conditions should not be considered in the scoring.
8. One object may involve more than one high-energy hazard (e.g., a suspended load often has gravity of the lifted object and the potential lateral motion of the object).
9. Observers must make reasonable efforts to verify that Direct Controls are installed and used properly.
10. The definitions of high-energy and Direct Control must be strictly applied.

Governance

As we have discussed, governance is an important part of collaboration. For example, EEI created a platform to govern HECA implementation in their sector. A community of practice maintains a formal HECA governance document[12]; collects, aggregates, and reports HECA data; creates community resources; and shares lessons learned. This governance has given leaders confidence that the method is applied consistently and reference data for interpreting their company trends. If you are in a different sector, I highly recommend working with peers to develop your own communities and governance structures and determining what works for your industry. Although HECA has a strict structure and set of rules, communities can create their own integration resources, methods of data analysis, dashboards, and benchmarking.

Trademark

Perhaps controversially, I trademarked HECA in 2024.[13] You may have noticed that in the title of this chapter. The purpose was not to restrict the use of HECA, but to preserve its integrity. The goal was to ensure that HECA remains a consistent, evidence-based standard that means the same thing to everyone. Just as TRIR has endured as a universal metric because of its uniform definition, HECA must be protected from drift if it is to serve as a viable alternative. If organizations begin modifying HECA by changing the definitions, rules, or thresholds, even slightly, we risk introducing confusion and weakening the collective value of the metric. The trademark offers a degree of legal protection against dilution, but more importantly, it reinforces our shared commitment to consistency. HECA is not a proprietary tool to be owned; it is a common language to be preserved so that we can finally measure what matters, together. If the metric is changed, it must be called something else.

STRENGTHS AND CHALLENGES OF HECA

A balanced understanding of any safety metric requires recognizing both its strengths and limitations. While HECA offers significant advantages over traditional metrics like TRIR, it is not without its challenges. The following features illustrate what sets HECA apart, highlighting its unique contributions.

Strengths

- ***Measures the safety of normal work*:** HECA measures in present tense, providing a method to objectively and practically answer, "How safe is this job?" This enables real-time monitoring and more agile business decisions.
- ***Directly measures safety as the presence of safeguards*:** While traditional metrics like TRIR measure the absence of incidents, HECA applies the concept that safety is the presence of safeguards.
- ***Focuses on SIF*:** By assessing high-energy hazards, HECA specifically targets conditions that could lead to serious injuries and fatalities (SIF).
- ***Reflects the understanding that error is normal*:** By focusing on Direct Controls, HECA accounts for the inevitability of human error, ensuring controls remain effective.

- ***Based on direct observation and strict definitions*:** HECA relies on the physics of energy and precise definitions, reducing bias when the rules are properly applied.
- ***Statistically stable and precise*:** Due to its design within routine site visits, HECA data are collected in large volumes, offering statistical stability far beyond traditional injury rates.
- ***Predictable and predictive*:** Early research shows that HECA scores can be predicted based on organizational factors and that HECA trends correlate with long-term SIF rates.[14]
- ***Supports learning*:** HECA reveals real-time trends, fostering proactive decision-making and investment in building capacity and resiliency rather than reacting to random noise.

Challenges

- ***Implementation complexity*:** HECA requires strict rules, detailed assessments, and significant investments in training and system integration.
- ***Resistance to change*:** Transitioning from traditional metrics to HECA involves a cultural shift that may face resistance from stakeholders.
- ***Data overload and analysis demands*:** HECA generates large datasets, which can overwhelm organizations lacking robust data management systems.
- ***Lack of Direct Controls for certain hazards*:** Feasible Direct Controls may not exist for all high-energy hazards, posing persistent challenges.
- ***Perception of imperfection*:** HECA's realistic scores may fall short of perfection, which can be challenging for organizations equating high scores with safety excellence.

BROADER REFLECTIONS ON THE CHALLENGES OF HECA

Throughout this chapter, I have obviously shown my bias in favor of HECA. Although it is built on a solid foundation, it remains as vulnerable to manipulation as any other metric. As the saying goes, "what gets measured gets manipulated," and HECA is no exception. Poor incentives can drive underreporting, misreporting, and other forms of manipulation, shifting the focus away from true safety. Without a clear sampling plan and reliable methods for calibration and validation, HECA risks becoming just another numbers game. To counteract this, companies must invest in training, calibration, and validation processes.

Training provides a consistent baseline understanding of how HECA should be applied; calibration exercises align observers to reduce variability; and validation through external and internal comparisons fosters trust in the data. All three techniques are crucial to applying HECA accurately and ensuring the trends and results can be trusted. For now, I recommend using HECA for learning, rather than comparing.

HECA is not a silver bullet. It measures one specific aspect of safety and must be complemented with other metrics to provide a complete picture. For example, organizations can explore how safety activities influence HECA trends or analyze how HECA correlates with injury rates. The future of safety measurement isn't about replacing TRIR with HECA, but rather creating a cohesive set of metrics that captures the entire safety system, from proactive monitoring to long-term outcomes.

CONCLUSION

Safety measurement is a complex challenge. Despite longstanding calls for alternatives, companies still rely on injury rates like TRIR. These metrics are used in fundamentally flawed ways, promoting reactive behavior, downplaying SIFs, and offering limited predictive value. Yet, they persist because viable, standardized alternatives have been elusive.

The EBS principles underpin HECA as a new alternative. By directly measuring the control of high-energy hazards, HECA is SIF-focused, measurable in real time, and grounded in rigorous analytical techniques. These strengths have made it a compelling option, gaining support across key industries such as utilities and infrastructure construction.

As safety metrics continue to evolve, HECA will play an integral role as a monitoring variable that links proactive safety activities (leading indicators) to injury trends (lagging indicators). This connection reveals what to change, how much to change, and what outcomes to expect, offering a clearer picture of safety performance over time. While a single metric can indicate status, using multiple complementary metrics tells a story, and stories drive better conversations that yield more informed decisions. By adopting balanced and interconnected safety metrics, organizations can shift from reactive firefighting to farming.

Moving forward, it is crucial to protect HECA as a standardized metric. Consistent implementation enables clear communication, aggregated data analysis, and community-driven innovation. While customization can be beneficial for safety programs, creating individual versions of HECA will undermine its value for both comparing and learning outside the company. Instead, organizations should share their ideas with the broader community, engage in HECA's governance, and work to preserve the method as a shared standard.

It will take time to get HECA right. Although leaders may be tempted to rush implementation, a full-scale rollout involves extensive training, calibration, and adaptation of data systems. Proper implementation typically takes a few years, not a few months. It is essential to give employees the time to pilot, refine, and fully adopt HECA, ensuring its success and impact on organizational safety. As the saying goes, "you can do it right, or you can do it again."

Although HECA enables organizations to learn from ongoing conditions whenever they choose, it doesn't eliminate the possibility of failure. When serious events do occur, whether near misses or actual injuries, we need a consistent way to classify them and extract learning. The next chapter will introduce the Safety Classification and Learning (SCL) model as a framework built on the same EBS principles used to underpin design, planning, monitoring, and measurement.

KEY TAKEAWAYS

1. Safety metrics shape how organizations set goals and measure progress toward those goals.
2. Using injury rates like TRIR to measure and reward safety performance is both philosophically and statistically flawed.
3. Underpinned by EBS and HOP principles, High-Energy Control Assessments (HECA) offers a new way to measure safety as the presence of safeguards by assessing normal work.
4. HECA assesses the control of high-energy during Safety Walks, offering a statistically valid measure of SIF-related conditions.
5. Consistent use of HECA enables the comparison and collaboration needed to make meaningful progress in SIF prevention.

NOTES

1 Kyung, M., Lee, S. J., Dancu, C., & Hong, O. (2023). Underreporting of workers' injuries or illnesses and contributing factors: A systematic review. *BMC Public Health*, 23(1), 558.

2 Conklin, T. (2016). *Pre-Accident Investigations: Better Questions-An Applied Approach to Operational Learning*. CRC Press, page 8.

3 Hallowell, M. R., Erkal, E. D., Sherratt, F., MacLean, B., & Davis, M. (2024). Safety performance measurement in environmental, social & governance frameworks. *Professional Safety*, 69(07).

4 This metric goes by different names and acronyms such as total recordable injury frequency (TRIF) and Total Recordable Injury and Fatality Rate (TRIFR). Globally, there are slight variations of this metric such as Lost Time Injury Frequency Rate (LTIFR), Lost Time Injury Incident Rate (LTIIR), and AFR (Accident Frequency Rate), which have regional differences in what must be recorded and reported and how the data are normalized (e.g., by 200,000 or 1 million worker-hours). Despite these differences, the basic structure of the metrics is consistent.

5 https://www.ncbi.nlm.nih.gov/pmc/articles/PMC10037763/.

6 Lingard, H., Hallowell, M., Salas, R., & Pirzadeh, P. (2017). Leading or lagging? Temporal analysis of safety indicators on a large infrastructure construction project. *Safety Science*, 91, 206–220.

7 The characteristics of the dataset and statistical tests are detailed in Salas, R. I. (2020). Empirical Evaluation of Construction Safety: Insights from Economic Trends, National Cultures and Workplace Injury Rates (Doctoral dissertation, University of Colorado at Boulder) and a corresponding journal paper Hallowell, M., Quashne, M., Salas, R., MacLean, B., & Quinn, E. (2021). The statistical invalidity of TRIR as a measure of safety performance. *Professional Safety*, 66(04), 28–34.

8 Erkal, E. D., & Hallowell, M. (2023). Moving beyond TRIR: Measuring & monitoring safety performance with high-energy control assessments. *Professional Safety*, 68(05), 26–35.

9 Oguz Erkal, E. D., Hallowell, M. R., Ghriss, A., & Bhandari, S. (2024). Predicting serious injury and fatality exposure using machine learning in construction projects. *Journal of Construction Engineering and Management*, 150(3), 04023169.

10 Edison Electric Institute. (2024). High energy control assessment (HECA) model. https://www.eei.org/-/media/Project/EEI/Documents/Issues-and-Policy/Power-to-Prevent-SIF/EEI-HECA.pdf.

11 Issued by the US Patent and Trademark Office (USPTO), registration number 7,373,532.
12 Edison Electric Institute. (2024). High energy control assessment (HECA) model. https://www.eei.org/-/media/Project/EEI/Documents/Issues-and-Policy/Power-to-Prevent-SIF/EEI-HECA.pdf.
13 Issued by the US Patent and Trademark Office (USPTO), registration number 7,373,532.
14 Oguz Erkal, E. D., Hallowell, M. R., Ghriss, A., & Bhandari, S. (2024). Predicting serious injury and fatality exposure using machine learning in construction projects. *Journal of Construction Engineering and Management*, 150(3), 04023169.

11 Safety Classification and Learning (SCL) Model

In the previous chapter, we explored how Energy-Based Safety (EBS) principles have infiltrated measurement, one of the most contentious topics in modern-day safety. We unpacked the dark history of safety measurement and explored how we can transition from measuring and reacting to statistical noise to monitoring SIF conditions via High-Energy Control Assessments (HECA). In this chapter, we turn our focus from monitoring conditions to learning from incidents.

Learning is the cornerstone of all progress, transforming our experiences into insight and our ideas into action. It is fundamental to human existence, enabling us to grow, adapt, and thrive in an ever-evolving world. Through continuous learning, societies advance, organizations innovate, and individuals achieve their fullest potential. When we learn, we not only expand our understanding of the world, but we also shape it.

The path to preventing SIFs is paved with deliberate and sustained organizational learning. Fundamentally, we must learn how to better control high-energy hazards, but how we learn depends on the topic and our role. For example, a new worker might learn how to identify hazards using the energy wheel, a leader may learn how work is performed by visiting sites, and the company may learn about high-energy exposures by trending observations. To really prevent SIFs, we must foster learning at all levels and embrace a multitude of discrete but interconnected roles.

This chapter will focus on how strict adherence to shared definitions can significantly expand the number of relevant learning opportunities for SIF prevention. We will explore how SIF-related incidents and observations can be objectively classified using EBS principles, how this enables multiple companies to aggregate their databases, and how properly contextualized information yields actionable insights. This new approach moves beyond traditional regulatory-driven classifications based solely on actual severity (e.g., lost-time injuries) to include potential serious injuries and fatalities (PSIFs) and condition observations, offering a richer foundation for learning and prevention.

THE SIF LEARNING PARADOX

Although SIFs are extraordinarily important, they are exceedingly rare. This creates a paradox where we urgently want to learn how to prevent SIFs, but they occur so infrequently that there isn't enough information to do so. This paradox often leads to a "whack-a-mole approach," where companies respond vigorously to each SIF, only to see new, seemingly unrelated incidents emerge later because the underlying patterns were unaddressed.

DOI: 10.1201/9781003641483-13

After a SIF, companies often rightfully feel compelled to demonstrate the seriousness of the incident and reaffirm their commitment to safety. This usually involves launching investigations aimed at preventing future occurrences. However, this approach can become perilous when findings from a small number of incidents are overgeneralized and addressed via sweeping corrective actions. For instance, retraining an entire workforce based on the actions of one individual may divert resources from far more effective interventions. Overgeneralization not only creates inefficiencies but also fails to tackle the real issues, resulting in a cycle where today's SIFs become tomorrow's clutter.

To make real progress in SIF prevention, we must work together to dramatically increase our capacity to learn. This can be achieved in three ways: (1) studying potential serious injuries and fatalities (PSIFs); (2) monitoring and analyzing ongoing conditions; and (3) sharing data across company boundaries. Achieving this goal requires precise definitions that are universally accepted and consistently applied. Such definitions enable a transition from isolated efforts by individual companies to a collaborative approach where we learn from a vast array of events and observations together.

DEFINITIONS ARE THE SCAFFOLD OF LEARNING

Although it may seem mundane, common definitions ensure precise communication and align our perception of what is relevant. Differences in definitions are the Achilles heel of shared learning because varying interpretations of SIFs and PSIFs derail communication and makes it impossible to share data. As discussed in Chapter 1, we made significant strides in defining SIFs as life-ending, life-threatening, and life-altering injuries – what we refer to as the LIFE model. However, defining PSIFs has proven more challenging due to the nuanced and complex nature of *potential* SIFs.

We are still in our infancy in terms of a safety vocabulary, and we have a lot to learn from other disciplines. Most definitions develop and mature over time, starting with personal interpretation (i.e., "you know it when you see it"), proceeding to lists (i.e., an extensional definition), and eventually to rule-based definition that describes the underlying phenomenon (i.e., an intensional definition). The definition of a planet provides an excellent example of this evolution.

CURIOUS CASE OF PLUTO

The term *planet* was first coined by the ancient Greeks to describe a celestial object that appeared to move independently from fixed stars. Later, in the 17th century, an extensional definition emerged with a comprehensive list of the nine observed planets. It wasn't until 2006 that the International Astronomical Union (IAU) created an intensional definition, classifying planets according to three rules: (1) it must be in orbit around the Sun, (2) it must be massive enough that it takes a nearly round shape, and (3) it must clear the neighborhood around its orbit. Using this rule-based definition, any newly observed celestial object may be consistently classified as a planet or not, and the findings can be unambiguously shared with the scientific community. This proved useful as our gaze into the cosmos has dramatically broadened.

The infamous consequence of this definition was exclusion of Pluto. Although Pluto did not change, it was "demoted" to a dwarf planet because it does not clear its neighborhood. No matter how many "bring Pluto back" t-shirts we make, it will not become a planet unless the scientific community agrees to change the definition. This is a mark of a mature scientific field.

The concept of PSIF is hardly new. For over 20 years, authors of white papers and editorials have made the case that PSIF events should be studied, but they did not offer a precise definition, which led to chaotic, inconsistent, and sometimes illogical approaches. In recent years, the concept of PSIF has rapidly evolved from individual interpretation to a rule-based definition. However, despite this progress, many organizations still default to lists. To demonstrate the value of a rule-based definition of PSIF, this section will discuss the problems associated with lists, and explain the motivation, process, and outcome of efforts to establish objective rules.

THE PROBLEM WITH LIST-BASED DEFINITIONS OF PSIF

The Campbell Institute[1] and others[2] made important early progress in the definition of PSIF by establishing lists like the one below that provide concrete examples of hazards that are PSIF-related. However, the lists still lead to inconsistency in classification and inclusion of far too many events (overclassification) because they fail to specify when something *is not a PSIF*. Without clear rules, any incident or condition can be classified as a PSIF, flooding company databases with hundreds of cases that dilute what is truly important. For example, overclassification stems from issues like:

- Subjectivity in the assessment of whether the condition had a sufficient danger to cause SIF harm (e.g., items 19 and 20);
- Generalized conditions that do not always have SIF potential (e.g., in item 1, a scratch to an arm does not have SIF potential just because it occurs in a confined space);
- "Other" categories (i.e., item 21) that can be applied broadly at the discretion of the analyst;
- No explicit consideration of the exposure to the hazard or the presence or absence of controls.

Example of a limited "PSIF" list based only on hazardous exposure

1. Confined Space
2. Lock-out tag-out (LOTO)
3. Work at height

4. Fall greater than 48 inches
5. Falling into deep water
6. Suspended load
7. Hot work
8. Arc flash
9. Fire
10. Explosion
11. Hazardous materials
12. Vehicle collision
13. Struck-by or caught between a vehicle or powered equipment
14. Contact with moving components of stationary machinery
15. Barricades or guarding has been defeated or bypassed
16. Contact with moving components of powered equipment
17. Pinched, caught between, struck by, or in the line of fire of a moving object with sufficient energy to cause SIF harm
18. Violent attacks by a person or animal species capable of inflicting SIF harm
19. Electrical contact of sufficient voltage/amperage to cause SIF harm
20. Uncontrolled energy sources like electrical, mechanical, hydraulic, pneumatic, chemical, thermal, high pressure, or potential energy
21. Any other SIF exposure situation not described

PURSUIT OF A RULE-BASED DEFINITION OF PSIF

In 2019, the Edison Electric Institute (EEI) commissioned a project to develop a rule-based PSIF definition, and I was fortunate to serve as the technical advisor. Our goal was clear: by defining PSIF and fostering collaborative learning, we hoped to make meaningful strides in reducing SIFs.

Our team of 20 senior safety professionals started with a period of exploration, where we aimed to identify and remedy sources of disagreement by reviewing controversial cases. One such case is provided as follows.

Controversial Case: A worker is at 20 feet of height and falls and was caught by a fall arrest system, which was designed and used properly. The worker was not injured.

In the review of this case, the team was completely split. Exactly 50% indicated that the event should be classified as a PSIF, and the other 50% indicated it should not. The major disagreement centered on the presence of the functioning lanyard. Some believed that the lanyard was a sufficient control because it was used properly, worked as designed, and eliminated SIF exposure. Others argued that the engagement of a fall arrest system categorically represented a PSIF. However, the team agreed on one point: the case did not represent success and highlighted the need for a category that falls outside of SIF, PSIF, or success. The consensus was that we must understand the current sources of disagreement in our classifications and strive for a model that drives consistency. If successful, we would be able to reach common conclusions, even for ambiguous and potentially divisive cases like the one above.

PSIF NOISE EXPERIMENT

We suspected that the statistical variability (noise) was not unique to our team. Therefore, to quantify the magnitude of this issue, we launched a PSIF noise experiment. We asked 113 practitioners to answer two simple questions for four test cases: *Is this a potential serious injury or fatality (yes or no)?* and *What is your confidence in your response (measured on a 1–5 scale, where 1 is extremely low and 5 is extremely high)?* Participants were encouraged to use any existing definitions or tools at their disposal.

The cases and results are shown in Table 11.1 below. The level of noise was astounding, with close to perfect disagreement on three of the four cases (note that 50% is perfect disagreement). Despite this high variability, the individual confidence was high (average of 4.6/5). Just like modern politics, most people disagreed but were confident they were correct.

THE SAFETY CLASSIFICATION AND LEARNING (SCL) MODEL

We used results of the noise experiment, the inventory of current practices, the literature review, and our recent research on EBS to draft the initial model. Our goal

TABLE 11.1
Results of the Noise Experiment

Case Example	Is this a PSIF? (% Yes)	Confidence (1–5 Scale)
Worker is at 70 feet of height with an 8 lb tool. There is no protection below, but work is under way below the elevated workspace. There is no lanyard on the tool and the tool has not been dropped.	68%	4.5
Worker is at 70 feet of height with an 8 lb tool. There is no protection below, but work is under way below the elevated workspace. The worker drops the tool, and it falls to the ground, but no one was injured.	93%	4.9
Worker is at 70 feet of height with an 8 lb tool. There is no protection below, but work is under way below the elevated workspace. The tool is dropped but is caught by a tool lanyard.	62%	4.5
Worker is at 70 feet of height with an 8 lb tool. The tool is dropped but is caught by a tool lanyard and there is a physical barrier in place to keep workers from entering the zone below the elevated workspace.	35%	4.4

wasn't to create a final product, but rather to build an initial framework that could be repeatedly tested, broken, and rebuilt until it yielded consistent classifications.

The process of "breaking" the model was fascinating. The team was asked to bring the most challenging, unusual, and interesting cases to the meetings. We examined cases that involved violent attacks by members of the public, explosions with far-flung debris, and super-hero reaction times, just to name a few. We also tackled simpler cases that sparked strong disagreement, like the one previously discussed. Our thinking was that if we could successfully classify the most challenging cases, the model would be strong enough to handle almost any case in the future.

The team discussions revealed two distinct ways of thinking that resulted in different conclusions. The first related to assessing the worst possible versus the most likely outcome. That is, while some focused on whether the event could have *possibly* caused a SIF, others judged whether the condition was *likely* to cause a SIF. Although seemingly inconsequential, these different views led to very different results. The team reconciled these perspectives concluding that we should not be asking the question, Does this incident have SIF potential? Because, as we discussed in Chapter 4, every hazard has the remote possibility of causing a SIF. Instead, we should ask: Is the most likely outcome of this event a SIF? The second major philosophical difference related to controls. Although some team members focused only on the seriousness of the hazard involved, others considered the presence and adequacy of relevant controls. After the discussions, the team concluded that cases where an adequate control was present were fundamentally different from those with an inadequate level of control. The team agreed that a new model must consider this perspective.

The final output was the Safety Classification and Learning (SCL) Model shown in Figure 11.1, which was originally published by EEI.[3] It might look complicated at first, but the model only asks four yes or no questions:

1. Was high-energy present?
2. Was there a high-energy incident?
3. Was there a Direct Control?
4. Was a serious injury or fatality (SIF) sustained?

The elegance is that these four simple questions can be used to classify any incident or observation, but the challenge is that the questions are not always easy to answer. Therefore, we will review exactly how to answer each question in the coming sections.

Question 1: Was High-Energy Present?

A flow chart should begin with a parent node that sets the direction for the entire analysis that follows. The influence of EBS on the SCL model is evident from the start, with the first question being related to high-energy.

This approach builds on the understanding that high-energy hazards are most likely to cause SIFs and low-energy hazards are not. Because the concept of PSIF revolves around potential outcomes, energy assessment is a perfect fit, providing an

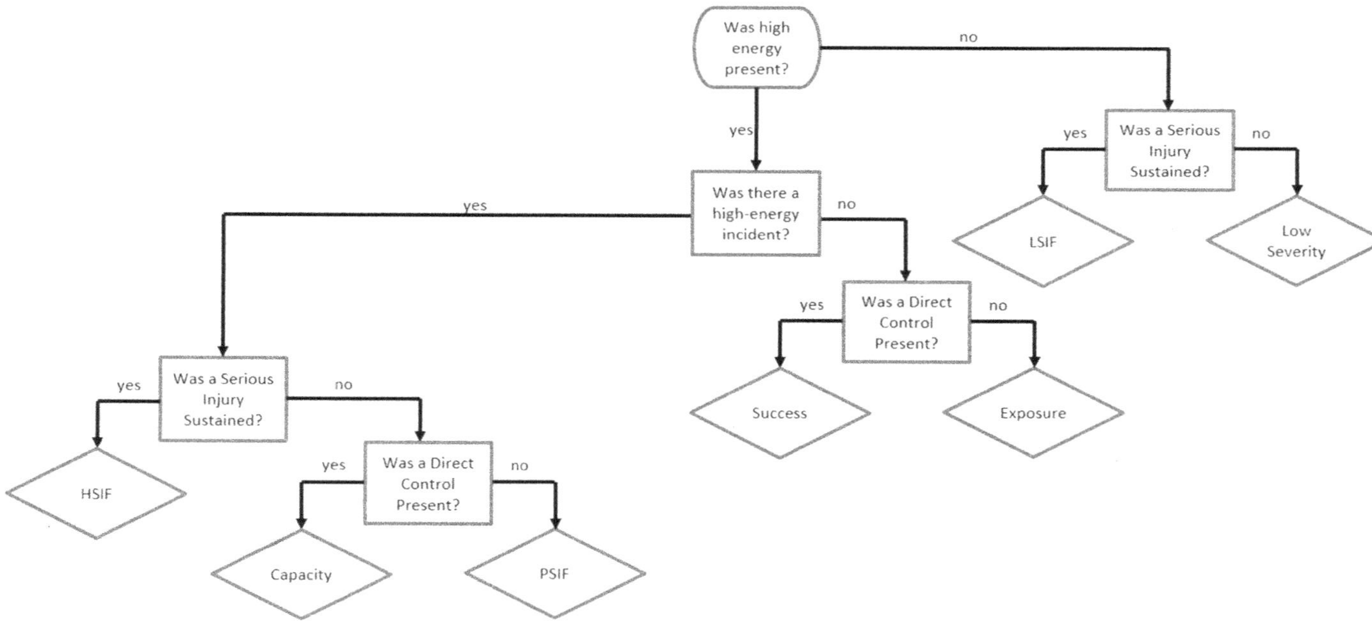

FIGURE 11.1 Safety classification and learning (SCL) model. This model classifies incidents and observations based on energy magnitude, the presence of controls, and the circumstances of energy release. It clarifies when a case qualifies as a PSIF and when it falls into another category.

This figure appeared in: Edison Electric Institute. (2024). Safety classification and learning (SCL) model. https://www.eei.org/-/media/Project/EEI/Documents/Issues-and-Policy/Power-to-Prevent-SIF/eeiSCLmodel.pdf. Reprinted with permission.

objective method to assess whether a hazard is serious. As a reminder, a high-energy hazard is an energy source that exceeds 1500 Joules.[4]

To assess energy magnitude for a hazard, we have two options: (1) use the high-energy icons shown in Figure 11.2 below, or (2) "do the math" using the equations in Appendix 2.

Question 2: Was There a High-Energy Incident?

If at least one high-energy hazard exists, then the next question is whether there was an "incident" related to that energy source. At first, we thought it would be easy to determine whether an incident had occurred or not. However, as cases were analyzed, it became apparent that assessing an incident can be nuanced and complicated.

The team settled on the definition of an incident as an ***instance where the high-energy source was released AND where the worker came in contact with OR proximity to the high-energy source***. This definition is graphically described by Figure 11.3.

To ensure consistent application of this definition, we defined energy *release* as an instance in time when there is a temporary loss of control of the high-energy, or when the high-energy changes state in the work environment. An example of a change in state is a tool that is dropped and transitions from potential to kinetic energy as it falls. An example of a temporary loss of control is a vehicle operator who falls asleep and departs from the established traffic pattern. In this context, loss of control typically refers to an instance when an operator temporarily loses command of a tool or piece of equipment that is related to the high-energy hazard. Note that some energy sources can be in a constant state of release such as flowing electrons in an exposed electrical system, hot surface temperatures, and radioactive material. The following example for electrical energy illustrates this point.

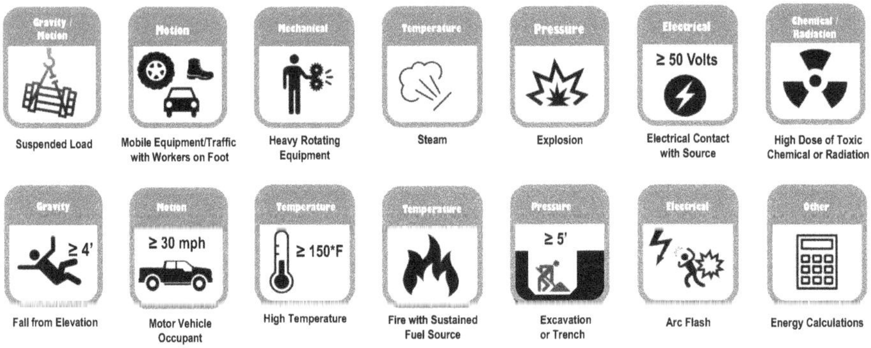

FIGURE 11.2 High-energy icons are examples of hazards that are almost always over 1500 Joules. These are a field-ready simplification of high-energy computations for hazard types that are almost always above the 1500 Joule threshold.

This figure appeared in: Edison Electric Institute. (2024). Safety classification and learning (SCL) model. https://www.eei.org/-/media/Project/EEI/Documents/Issues-and-Policy/Power-to-Prevent-SIF/eeiSCLmodel.pdf. Reprinted with permission.

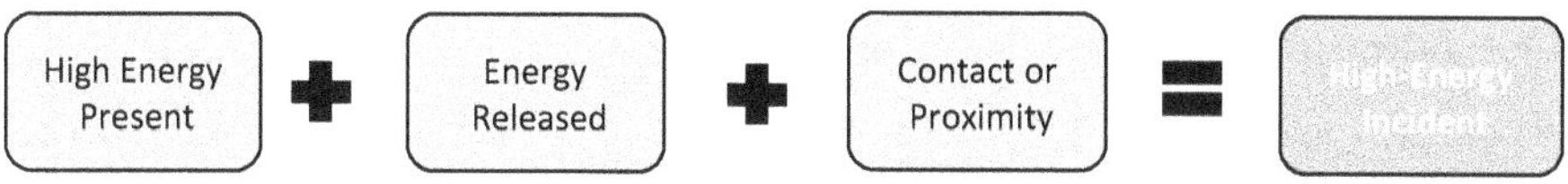

FIGURE 11.3 Components of the high-energy incident definition. A high-energy incident is defined by three elements: the presence of high-energy, a release of that energy, and either physical contact or close proximity. All three must be present for an event to meet the definition.

This figure appeared in: Edison Electric Institute. (2024). Safety classification and learning (SCL) model. https://www.eei.org/-/media/Project/EEI/Documents/Issues-and-Policy/Power-to-Prevent-SIF/eeiSCLmodel.pdf. Reprinted with permission.

IS THERE AN ENERGY RELEASE?

✓ The soil wall of a trench collapsed when an excavator passed by.

✖ An employee was observed working at height without a fall arrest system and was called down to speak to a supervisor.

To be an incident, the worker(s) must also either have contact with the energy or be in proximity to the energy. *Contact* is defined as an instance when the high-energy is transmitted to the human body. *Proximity* is defined as a hazardous circumstance where the boundary of the high-energy exposure is (1) within 6 feet of the worker; (2) within a confined space; or (3) within a situation where a worker cannot escape the energy source (restricted egress). The 6-foot threshold was selected because it approximates the radius of influence of a person (6-feet height and 6-feet wingspan).

Interpretation of proximity can sometimes be challenging. For example, we once discussed a case where an individual observed a falling object and, because of the time it took the object to fall, was able to move out of the way before it crashed to the ground. For such cases, we decided that proximity includes the entire path of travel and the individual's closest point to that path. In this case, the employee was lucky to get out of the way, but we determined there was proximity because they were in the path of travel when the object was initially dropped.

IS THERE PROXIMITY?

✓ A water truck loses its brakes and careens into a ravine. The employee was able to jump free from the truck as it was initially moving.

✖ A trench wall collapses but the crew was not in the trench at the time because they were on lunch break.

Question 3: Was a Direct Control Present?

A core principle in the SCL Model is that the primary differentiator between success and failure is the presence or absence of adequate controls. Consistent with Chapter 5, the team defined adequacy as a Direct Control that (1) is specifically targeted to the high-energy source; (2) effectively mitigates exposure to the high-energy source when installed, verified, and used properly (i.e., a SIF should not reasonably occur if these conditions are present); and (3) is effective even if there is unintentional human error during the work (unrelated to the installation of the control).

Although we have since introduced the concept of Alternative Controls, this remains a binary decision in the SCL model: Was there a Direct Control or not? We have maintained this distinction because we still aspire to have a Direct Control for every high-energy hazard and the SCL classifications are meant to enable learning about how this may be achieved.

Question 4: Was a Serious Injury Sustained?

When determining whether an injury is serious, refer to the LIFE model and the associated SIF criteria in Chapter 1 and Appendix 1, respectively. These provide strict guidance about what constitutes a serious injury and examples of common injuries that conform to that definition, respectively. Note that this question is whether there was an *actual* SIF, and not whether there was or could have been a *potential* SIF.

After creating the SCL model, we conducted another noise experiment with 25 cases and another 70 industry professionals. The result: 94% agreement even for the most challenging cases.[5] This was a massive improvement over the near perfect disagreement observed previously.

SCL MODEL CATEGORIES

The SCL Model can yield one of seven possible outcomes: HSIF, LSIF, PSIF, Capacity, Exposure, Success, and Low-Severity. A definition and interpretation of each of these classifications are provided below, consistent with the four SCL model questions. For each classification, a basic example is provided. Three examples are provided in Appendix 4.

High-Energy Serious Injury or Fatality (HSIF): Incident with a release of high-energy where a serious injury is sustained. These are high-priority events because a worker, their family, co-workers, and the organization are all deeply affected. The organization must respond seriously to such events and seek to learn to prevent future failures. In almost every HSIF event, a Direct Control was absent or failed.

> **HSIF Example**: An employee was paralyzed after being trapped at the bottom of a 10-foot (3-meter) trench collapse.

Low-Energy Serious Injury or Fatality (LSIF): Incident with a release of low energy where a serious injury is sustained. Typically, LSIF incidents are related to health and physiology. Unlike HSIF events that mainly relate to engineering controls,

LSIF events are typically best addressed by an industrial hygienist or a medical professional. Thus, the competencies needed to learn and the means of preventing future incidents may require consultation outside of the safety profession. LSIF are not less important than HSIF, they are simply different.

> **LSIF Example**: An employee went into a coma after contracting West Nile virus from a mosquito bite.

Potential Serious Injury or Fatality (PSIF): Incident with a release of high-energy in the absence of a Direct Control where a serious injury is not sustained. PSIF incidents have the same circumstances and characteristics as HSIF events except for the SIF outcome. In other words, we were simply lucky that a SIF did not occur. These events are great chances to learn because no one was hurt, and the people involved can join the learning process.

> **PSIF Example**: An employee was standing within 3 feet (1 meter) of a large steel I-beam that was dropped from a height of 10 feet (3 meters) during lifting operations. No injuries occurred.

Capacity: Incident with a release of high-energy in the presence of a Direct Control where a serious injury is not sustained. Unlike PSIF cases, Direct Control(s) are present in Capacity cases. Because a high-energy incident occurred in the presence of a Direct Control, capacity incidents are not inherently positive or negative. Rather, they represent excellent learning opportunities because they provide insight on what triggered the energy release and the opportunity to verify the resilience of their controls without negative consequences.

> **Capacity Example**: A fire ignited in a fuel storage area, but the automated fire suppression system activated immediately, extinguishing the flames before they could spread or harm nearby workers.

Exposure: Condition where a high-energy hazard does not have a corresponding Direct Control. *Unlike* incidents, an exposure is an observable condition. Exposure conditions are the same as PSIF and HSIF except that an incident hasn't happened yet. Thus, learning can occur from exposure events before a negative incident occurs. Observations of conditions can also be made regularly, resulting in a higher volume of learning opportunities.

> **Exposure Example**: Employees were observed working within 6 feet (2 meters) of an energized 13.8 kV switchgear cabinet with the doors open, and no energy isolation measures in place.

Success: Condition where a high-energy hazard does have a corresponding Direct Control. Interestingly, the SCL Model creates a clear definition of what success looks like for SIF prevention. Here, success is distinguished from all other observations by the presence of Direct Controls. As with exposure cases, success cases are conditions that may be regularly observed. Therefore, they can be studied in high volume.

Success Example: During maintenance on a high-pressure steam line, the line was fully depressurized, isolated with double block and bleed valves, and verified safe before work began.

***Low-Severity*:** These are situations that did not result in, and are unlikely to result in, a SIF. While they may not be directly relevant for SIF prevention, low-severity incidents still play a role in overall safety management. They help identify trends, improve work practices, and reinforce a culture of hazard awareness. Low-energy hazards should be addressed through general injury prevention programs rather than formal control analysis, ensuring that they are effectively managed without detracting from SIF-focused efforts.

Low-Severity Example: A loose electrical cord was observed across a walkway, creating a trip hazard.

MULTIPLE-ENERGY CASES

About 20% of cases involve more than one high-energy energy hazard.[6] We have seen falls within proximity to electrical equipment; heavy equipment operating pinning a worker against an unsupported excavation wall; and suspended loads operating over high-pressure pipelines. When more than one high-energy hazard is present, the SCL model should be applied separately for each hazard, which may result in multiple classifications and multiple learning opportunities for one case.

Take the following example: A worker fell from a 20 ft platform and was caught by a fall arrest system that was installed, verified, and used properly. After the fall, they were left dangling within 2 feet of an unprotected live electrical transmission line. No injury was sustained, and the worker was lowered within 2 minutes. In this case, we would arrive at Capacity for fall to height over 4 ft hazard, and PSIF for the high-voltage electrical hazard. Although there is only one case, there are two high-energy hazards, two classifications, and two potential learning opportunities.

GOVERNANCE AND SHARED LEARNING

Once the SCL model was launched, we quickly saw the need for governance. There was an imminent risk that companies would begin to customize and adapt the SCL Model to align with their individual philosophies, which would quickly erode our ability to collaborate. To manage this risk, we created a community of practice

(COP) to support both governance and learning. Like the IAU governs the definition of a planet, our COP governs the definitions in the SCL model. Any member may propose a change to the model, which is followed by a meeting where the COP listens, discusses, and votes.

To quote George E.P. Box, "All models are wrong, but some are useful." The corollary is that a model's utility is determined by how often it is used and whether it provides valuable insights. The greatest testaments to the SCL model are the rate at which it is being used and the community that has formed around it. Creating or participating in communities of practice related to your specific industry sector can be immensely valuable.

As of 2024, the community submits over 10,000 SCL cases annually and uses a live dashboard to aggregate, inspect, and analyze the data.[7] Historically, groups like this would learn from about a dozen actual SIFs each year. With a thousand times more SIF-related data, we're now seeing patterns we would have missed before. Below are five examples of key learnings:

1. Heavy mobile equipment with workers on foot is the most common exposure because controls like spotters are prone to human error.
2. Falls into trenches are difficult to control because of the challenges in securing barriers to soft ground during dynamic work.
3. Most motor vehicle incidents are classified as Capacity events from the occupant's perspective, thanks to modern vehicle controls like seat belts, airbags, and reinforced frames.
4. Flaggers in work zones are almost always exposed to uncontrolled high-energy because Direct Controls for adjacent traffic are usually not economically feasible (ironically, they are sometimes listed as the control for the other workers in the work zone).
5. About a third of high-energy hazards rely on Alternative Controls rather than Direct Controls.

These insights and others have led to industry-level learning teams that are working to address these exposures, involving front-line workers, supervisors, clients, contractors, suppliers, equipment manufacturers, and technology companies.

Many different analyses and figures can be created from SCL Model data. A common example is shown in Figure 11.4, which plots the proportion of high-energy hazards with a corresponding Direct Control for both industry averages and an individual company. Such data reveals which high-energy hazards are commonly controlled in the industry and which are not (represented by the bars in the chart). By comparing against the industry trends, a company (represented by the dots) can identify where they are stronger and weaker than the industry average. For example, the company can see from Figure 11.4 that they are comparatively weak at controlling suspended loads, indicating that they may have something to learn from their peers. Alternatively, they are comparatively strong at controlling electrical contact and arc flash and may have something to share with their peers. Such insights are often unique to individual companies and can instigate SIF-related learning that was derived from no-consequence incidents and regular observations. We have found that every company has something to share and something to learn.

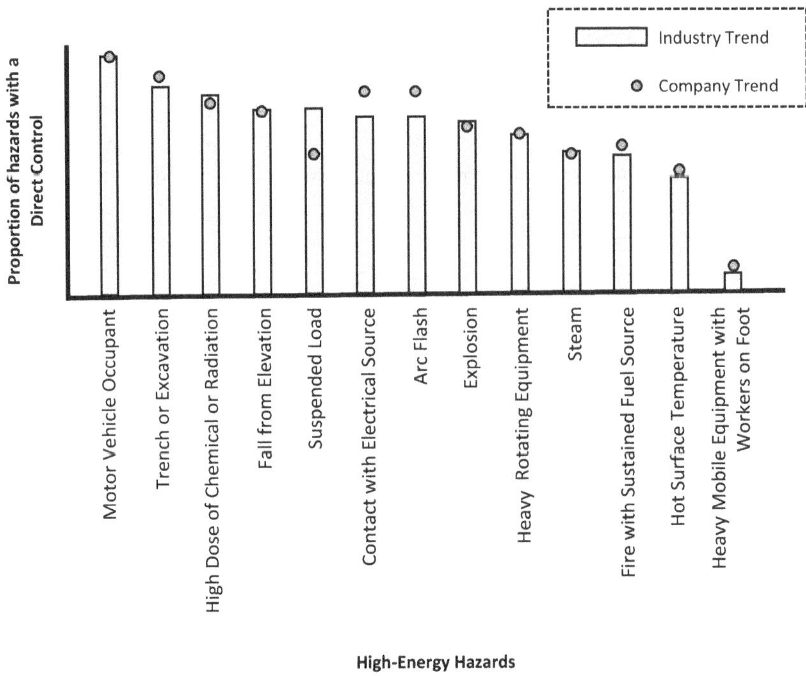

FIGURE 11.4 Hypothetical comparison of industry-level and company-level trends. This figure compares the proportion of high-energy hazards with Direct Controls between a company and the industry average, helping identify areas where the company can learn from others and where it may be able to teach others.

LESSONS LEARNED AND RECOMMENDATIONS

After the first 5 years with the SCL model, we documented a series of lessons learned, which have proven helpful to those beginning to use the SCL Model.

1. **Follow the SCL Model exactly and answer all questions before classifying**. One of the most common sources of confusion and misclassification occurs when questions are skipped, or assumptions are made instead of collecting all the evidence. All four questions (and their sub-questions) should be answered carefully and completely before a classification is made.
2. **Avoid "what-if" scenarios**. When beginning to use the SCL Model, analysts tend to consider how changes in hazard circumstances and proximity would affect the classification of the event. Although interesting, "what-if" scenarios should not be considered in the formal classification of the event or observation. Instead, consider only the facts and avoid fabrication or speculation. *Nothing other than the facts should drive the classification; we must not fall prey to our feelings and speculation.*
3. **Provide targeted training and the opportunity to calibrate**. Effective use of the SCL Model requires both training and practice. Specifically, analysts

should be trained on how to estimate energy magnitude, evaluate incidents, and assess controls. We have found that learning is accelerated, and performance improves when challenging cases are reviewed as a group. Several companies have seen value in having both widespread training across the safety team and deeper expertise for a smaller group of internal subject matter experts. This calibration is essential for consistent application.

4. **Different classifications can spur different learning**. Although we initially set out to define a PSIF, we found that it was best defined in the context of other learning opportunities. For example, the SCL model provides an important distinction between high-energy events where we are lucky (PSIF) and where we were prepared (Capacity). These different types of events can be used to direct different types of learning. For example, since research unequivocally shows that the causes of PSIFs are the same as the causes of actual SIFs,[8] we can study PSIFs as a proxy. Alternatively, Capacity incidents allow us to test the reliability and limitations of controls during an energy release. Each classification offers something different to learn, and no one type is more important than the others.
5. **Decouple SCL classification and learning effort**. Some organizations have developed an investigation program that aligns with the SCL categories. For example, deep investigations may be deployed for HSIF and PSIF events and lower levels of investigation may be deployed for capacity or exposure events. However, despite their immense impact, some HSIF cases are relatively weak learning opportunities, while some Capacity and Exposure events are relatively rich. Therefore, we recommend maintaining a degree of flexibility in how cases are investigated, directing resources to cases that are most interesting and useful.
6. **SCL provides a framework that supports learning from success**. An emerging soundbite among safety professionals is that we should learn from success, but this idea is often treated as an afterthought. The SCL model brings structure to this concept, turning an abstract idea into a tangible effort. Now that we have a precise definition of success (i.e., a condition where a high-energy hazard has a corresponding Direct Control), we can systematically study why these controls are installed, verified, and used properly. Investigating success allows us to uncover the conditions, decisions, and behaviors that enable strong controls, providing a blueprint for replicating these outcomes across similar work. By contrasting the factors that lead to successes with those that result in exposures, we open the door to being truly predictive. This represents a massive, untapped opportunity to shift safety from a reactive to a proactive, investing more energy in amplifying what works rather than merely avoiding what doesn't.
7. **PSIF and Capacity cases may encourage companies to reallocate attention and resources previously focused on low-severity incidents**. In a safety system that prioritizes SIF prevention, these events warrant significantly more focus than low-energy or minor recordable injuries that historically drew leadership attention due to their influence on injury rates. The SCL model, with its formal structure and strong community support, helps organizations refocus on what matters most for SIF prevention.

WHAT ABOUT MEASURING PSIF RATES?

Measuring PSIF rates may seem like a logical progression for companies focusing on SIF prevention and transitioning away from traditional metrics like injury rates. However, measuring and tracking PSIF rates is fundamentally flawed for several reasons.

Firstly, the reliability of PSIF data is inherently questionable. Potential incidents are "hidable" when there are no witnesses, and they do not result in visible damage or injury. Without psychological safety (i.e., trust), a company will only learn about hidable incidents if the employees are willing to share. For example, consider a scenario where a crew drops a heavy load with no damage, injury, or witness. If the crew chooses to clean up without reporting the event, the company will remain oblivious. The reasons why employees do not report incidents include fear of retaliation, concern over missing a PSIF target, or the desire to avoid the hassle of reporting.

Since selective reporting happens in every organization, PSIF rates lack unidirectional clarity. A high PSIF rate could indicate a genuinely unsafe environment, or it could reflect a robust reporting culture. Conversely, a low PSIF rate might suggest effective control of risks, or it could signal a weak reporting culture. Because of this ambiguity, PSIF rates are not indicative of performance.

The emphasis on achieving low PSIF rates can also create perverse incentives. If a company's goal is to minimize its PSIF rate, employees might underreport incidents just to meet this target, exacerbating the very risks the metric is intended to mitigate. Put simply, if you want low PSIF rates and you let your employees know it, that's exactly what you'll get, even is nothing actually improves.

CONCLUSIONS

Many safety professionals share the same vision: Eliminating SIFs. Achieving this goal requires continuous learning and improvement that extends beyond company boundaries. The safety community is addressing this issue by expanding learning opportunities through PSIF reporting and fostering collaborative learning that is grounded in the rule-based definitions of the SCL Model. As the old African proverb goes, "If you want to go fast, go alone; if you want to go far, go together." The SCL model has been an excellent vehicle to help us "go together."

Although the SCL model serves as an effective framework, it must be integrated within an environment where individuals feel safe to share ideas, admit mistakes, and question practices without fear of retribution. This trust encourages open communication and ensures that employees are comfortable sharing incidents and other information that they could easily hide.

With improved organizational culture, a rigorous classification model, and a collaborative community, we have begun to uncover previously hidden trends. For example, we are answering critical questions such as: Which high-energy hazards are well controlled, and which are not? Which controls are being installed, verified, and used properly, and which are not? With tens of thousands of cases submitted, these trends are statistically meaningful and provide far greater insight than we could ever have gleaned from SIFs alone.

KEY TAKEAWAYS

1. PSIF events offer significant learning potential, but inconsistent classification methods create noise and limit their utility.
2. The SCL Model, grounded in energy-based safety and controls analysis, was developed to address this challenge.
3. An integrated team of academics and industry professionals collaborated to design the SCL Model for practical application.
4. A community of practice supports SCL Model implementation through calibration, data sharing, and advocacy.
5. The SCL Model enables organizations to shift focus from low-severity incidents to meaningful SIF prevention efforts.

NOTES

1 Campbell Institute. (2018). Serious injury and fatality prevention: Perspectives and practices. Available at: https://www.thecampbellinstitute.org/wp-content/uploads/2018/11/9000013466_CI_Serious-Injury-and-Fatality-Prevention_WP_FNL_single_UPDATED-11-14-18_optimized.pdf.

2 Martin, D. K., & Black, A. (2015). Preventing serious injuries and fatalities: A new study reveals precursors and paradigms. Available at: https://dekra-insight.com/images/white-paper-documents/wp_preventing-sif_us_A4.pdf.

3 Edison Electric Institute. (2024). Safety classification and learning (SCL) model. https://www.eei.org/-/media/Project/EEI/Documents/Issues-and-Policy/Power-to-Prevent-SIF/eeiSCLmodel.pdf.

4 Note that EEI deferred to a more conservative 500 ft-lbs of energy as the threshold to be more inclusive and to communicate the method in companies accustomed to Imperial units. For non-EEI members, I suggested using 1500 Joules as the threshold to be consistent with the research.

5 Hallowell, M. R., & Spencer, C. (2024). Safety classification & learning model: Defining & classifying potential serious injuries & fatalities. *Professional Safety*, 69(1), 18–26.

6 Hallowell, M. R., & Spencer, C. (2024). Safety classification & learning model: Defining & classifying potential serious injuries & fatalities. *Professional Safety*, 69(1), 18–26.

7 To learn more about this community of practice, visit www.powertopreventsif.com.

8 Bayona, A., Hallowell, M. R., & Bhandari, S. (2024). The things that hurt people are not the same as the things that kill people: Key differences in the proximal causes of low-and high-severity construction injuries. *Journal of Construction Engineering and Management*, 150(8), 04024089.

12 Ten Steps for EBS Integration and Ten Lessons Learned

The safety profession stands at a crossroads. Despite decades of progress in reducing less severe incidents, SIFs have plateaued. This is not merely a statistical blip; it reveals a deeper truth: SIFs occur for different reasons and require a different kind of attention. This explains why even organizations with robust safety systems and low injury rates still have SIFs. It also raises the question whether generalized approaches to safety are truly capable of preventing the most devastating outcomes.

Most practitioners understand the urgent need to address SIFs, but targeted and practical solutions have been challenging. The promise of this book was a scientific approach to SIF prevention. EBS delivers with a set of first principles that enable organizations to identify and address SIF-related conditions before they cause irreparable harm. Just as Newton's laws provide a framework to understand and interact with the physical world, Energy-Based Safety (EBS) offers a lens through which we can examine and control the forces that threaten workers' lives.

This book introduced four EBS principles that underpin SIF prevention: energy causes harm, more energy causes more harm, controls save lives, and people drive performance. Using existing safety programs as a scaffold, EBS sharpens focus on SIF prevention and can be used as a framework to align activities, roles, objectives, and metrics to this objective. This approach is akin to the Trellis and the Vine, an analogy that I picked up from my friend Jack Suehiro.

The Trellis and the Vine offers an analogy for the integration of EBS into an existing safety system. As a trellis provides structure and direction that guides a vine's growth with purpose and strength, a well-designed safety program establishes the framework needed to integrate EBS principles. Key activities such as thoughtful design, pre-work planning, Safety Walks, monitoring, measurement, and learning form the scaffolding that supports an organization's growth toward SIF prevention. Without this framework, efforts can become disorganized and ineffective, much like a vine sprawling without direction.

DOI: 10.1201/9781003641483-14

> For the vine to thrive, the trellis alone is not enough. Organizations must have trust, commitment, and resources that serve as the nutrients that allow the EBS principles to take root. Without these essential elements, it is like planting in sand – growth would be stunted, and the seeds may never reach their potential. Similarly, leadership provides the sunlight, energizing and sustaining progress, ensuring the vine grows with strength and resilience. Together, these elements work in harmony to create a system that not only prevents SIFs but ensures a thriving, interconnected network of safeguards capable of reaching every corner of the organization.

In this final chapter, we will touch upon what we have learned about sequencing, timelines, and how to overcome common pitfalls. This topic could be a book of its own. To keep this chapter focused, I've distilled the key insights into ten integration steps and ten lessons learned.

TEN STEPS FOR EBS INTEGRATION

Every organization's journey with EBS will be different. The approach will be influenced by factors such as the structure and maturity of the organization's current safety program; its regulatory environment; and demands from stakeholders like board members, executives, and clients. However, despite these differences, we have found the following ten-step process to be generally effective for most organizations.

Step 1: Secure Leadership Buy-In

Garnering strong support from senior leaders is an essential starting point because leaders control goals, incentives, and resources. Without leadership support, the organization will not be equipped to drive systematic change. For most senior leaders, transitioning to a SIF-focused strategy is not trivial, as it may challenge conventional paradigms such as the safety pyramid, TRIR, and "zero-injury" goals. To accept that a new approach is needed, leaders must first be deeply concerned by the SIF plateau and understand the drawbacks of antiquated approaches. Success in these discussions is signaled when leaders ask questions like: Why have we gotten to this point? and What should we do differently? These questions set the stage to pose EBS as an evidence-based solution.

Step 2: Conduct a Readiness Review

Before integrating EBS, organizations must first assess their readiness through a structured review of their safety practices. This includes evaluating whether they have the necessary knowledge, personnel, processes, and systems to support successful implementation. As discussed throughout this book, EBS principles must be embedded in core safety activities such as pre-job briefs and Safety Walks. For instance, a company cannot effectively adopt the energy wheel without first establishing strong pre-job meetings where crews actively engage in hazard discussions. Without that foundation, the energy wheel will have limited impact.

Before EBS concepts can be introduced to the field, the safety function itself must be ready to lead and support. As integration progresses, safety professionals should stay at least one step ahead, building the expertise and capacity needed to guide each stage of the transition.

Step 3: Pilot with a Small Group

It may be tempting to train everyone all at once. However, a pilot provides an opportunity to test the effectiveness of EBS strategies in a small group before they are introduced more broadly. I recommend picking a project, region, or facility that is respected throughout the company and working with them to introduce the energy wheel first. Let them know that you want their feedback on how the initiative should be introduced more broadly, giving them a chance to help design the integration strategy. Pilots help create advocates, ensure the approach is well-aligned with the needs of the workforce, and give confidence that the initiative will deliver on its promises. They also align with the "fail fast" philosophy, allowing any issues to be identified and corrected before full integration.

Step 4: Develop Supporting Resources

Before large-scale roll-out of any phase, it is important to create practical resources that reinforce EBS principles and make them easier to understand and use. Visual and tactile tools make concepts like the energy wheel and high-energy icons accessible and memorable. Resources can take many forms, such as hard hat stickers, toolbox magnets, posters, videos, or even creative items like branded lunch boxes or cornhole games. The goal is to create clear, practical reminders that encourage regular use of EBS tools without crossing the line into slogans or propaganda.

In addition to visual resources, functional tools like forms and templates should be designed to align with EBS principles. Pre-job safety briefs, observation forms, audit checklists, and incident reports should be structured to collect the type of information that supports EBS integration. For example, forms might include fields to identify high-energy hazards, document the presence of Direct and Alternative Controls, or evaluate whether appropriate energy mitigation strategies are in place. Aligning both visual and functional resources with EBS creates a cohesive system that embeds these principles into how the work is performed.

Step 5: Train Field Leaders First

The concept of a train-the-leader program builds on the well-known idea of train-the-trainer but takes a distinctly different approach. Unlike train-the-trainer programs that prepare individuals to teach others, train-the-leader programs equip field leaders with a deep understanding of EBS principles so they can lead by example. This ensures that leaders are not just delivering information but actively demonstrating, applying, and championing EBS in their daily work. When they fully understand its value and integrate it into daily operations, their teams naturally follow suit. In short, field leaders must begin to "show up differently," with a visible commitment to preventing SIFs that inspires front-line workers and with the depth of understanding needed to lead them effectively.

Step 6: Engage and Train Front-Line Workers

Training is critical to ensure that EBS principles are understood and consistently applied. EBS doesn't require big system changes, but people do need to understand it and feel confident using it. This competency hinges on clear and consistent training. Video training is an effective way to scale this effort, providing a uniform foundation and preventing the miscommunication that can occur when multiple trainers deliver the message locally. However, video alone is not enough. This is where a train-the-leader approach becomes essential. Supervisors and field leaders who have a deep understanding of EBS principles can lead follow-up discussions and planning sessions, reinforce key concepts, answer questions, and ensure EBS is being applied as intended. By combining consistent messaging from video training with supervisor-led reinforcement, this two-pronged approach maintains clarity while allowing flexibility to address workforce-specific needs.

Step 7: Integrate EBS across Safety Activities

Building on the success of training and the competency developed, the next step is to embed EBS into daily safety activities. Once workers understand how to apply EBS concepts, that knowledge must be integrated into existing safety practices at every level of the organization. Although pre-job briefs are an effective starting point, EBS principles should extend far beyond that one activity to avoid the perception that EBS is only for the field. For example, safety professionals and managers can apply EBS during Safety Walks to reinforce the control of high energy as a priority; designers can apply EBS during safety in design reviews to eliminate high-energy hazards before they reach the field; and leaders can reference EBS concepts in safety conversations with front-line employees during their field engagements. If front-line workers are expected to apply EBS in pre-job briefs, they should see those same principles reflected in other safety activities, corporate messaging, and even metrics.

Step 8: Measure and Adapt the Safety System

Measuring the impact of EBS is essential for understanding its effectiveness and identifying areas for improvement. A good place to start is by collecting baseline data that serves as a reference for measuring progress. This begins with identifying the core competencies that EBS should enhance, such as conducting more complete hazard scans, prioritizing and controlling the most serious hazards, and having more focused safety discussions. Baseline data can be gathered by observing pre-task planning, reviewing safety documents, and conducting Safety Walks before EBS integration. With these data as a reference, improvements can be measured directly. Qualitative insights add depth by answering key questions: Are safety conversations becoming more focused? Are workers and leaders identifying and addressing gaps in controls? These behavioral shifts signal that EBS is taking root. Since EBS is an iterative process, changes will likely be needed along the way. These adaptations are a hallmark of growth.

Step 9: Sustain

EBS is not a "plant-and-go" effort. Like a vine on a trellis, it requires ongoing reinforcement, retraining, and refinement to remain effective. Without deliberate effort, even the best programs fade. In addition to including EBS in onboarding training for new employees, existing employees should continue to be given learning opportunities. For example, regular refresher training, live safety demonstrations, and hands-on coaching keep EBS principles fresh and ensure consistent application. Demonstrations on real-world hazards help workers connect EBS to practical safety measures; coaching on tools like the energy wheel or HECA reinforces correct usage and ensures continuous learning; and aligning incentives and recognition with effective integration encourages employees to stay engaged with the SIF prevention effort. By continuously retraining, refreshing, refining, and rewarding, organizations ensure EBS stays relevant and impactful over time.

Step 10: Collaborate

SIF prevention depends on collaboration and shared learning beyond company boundaries. Therefore, the final step in EBS integration is to engage with the broader safety community to exchange best practices, benchmark progress, and collectively advance toward SIF elimination. Sharing lessons learned not only strengthens internal efforts but also contributes to a larger movement. By participating in cross-industry discussions, research initiatives, and collaborative learning networks, companies ensure that EBS continues to evolve and improve. This step reinforces the idea that SIF prevention is a shared responsibility, not a competitive advantage.

TEN LESSONS LEARNED (OFTEN THE HARD WAY)

Step 10 emphasized that no company can eliminate SIFs alone. Progress depends on shared learning across industries. Through our collaborations, we have seen both successes and struggles in EBS integration, and the lessons learned from these experiences may help you avoid common pitfalls. While each organization's journey will be unique, these insights provide a roadmap for overcoming challenges and ensuring EBS delivers lasting change.

Lesson 1: Brace for Pivotal Moments

At some point, leadership will face defining tests that determine if EBS is a lasting transformation or a short-lived initiative. One of these moments comes with the inevitable spike in recordable injuries. The pressure to react will be immense. Will they stay the course with SIF prevention or revert to chasing low-severity events? If leaders panic and abandon EBS, they will reinforce the idea that safety is about numbers rather than actual risk reduction. The best organizations prepare for this moment in advance. Before launching EBS, leaders should define what success looks like beyond traditional lagging indicators and commit to measuring safety by the presence of strong controls, not just the absence of injuries. A recordable spike is not a failure. It is natural and expected statistical noise, and a critical part of progress in SIF prevention.

Another pivotal moment comes when enthusiasm fades. Early on, engagement is high as EBS feels fresh and promising. But after a few months, interest often dips as challenges emerge, other priorities compete for attention, and the imperfections of EBS become apparent. This is natural. Organizations should anticipate this lull from the outset and have a plan to push through. Success is not about maintaining early enthusiasm but about working through the lulls, refining the approach, and embedding EBS into everyday work. Those who persevere through these pressures will be the ones who create real and lasting change.

Lesson 2: Lean into the Evidence

EBS is not a trend or a passing safety initiative, it is built on evidence. Organizations that ground their safety programs in scientific methods build credibility and engagement at every level. Workers appreciate understanding why EBS works, from how the brain processes hazards to why energy is the key factor in SIFs. Leaders who clearly communicate the research behind EBS help their teams see it as a structured, research-backed approach rather than just another corporate initiative. Safety programs that rely on slogans and intuition instead of hard data lose momentum over time. The most successful companies stick to evidence-based methods that are guided by research, not assumptions.

Lesson 3: Customize the Integration, Not the Science

EBS works best when its scientific foundation remains intact. Core elements like the energy categories, the 1500-Joule threshold for high-energy, and the rules for Direct Controls should not be customized. If every company modifies them, EBS is no longer an evidence-based strategy because it loses its scientific foundation. Just as a common standard allows different languages to be translated, the EBS principles act as a kind of Rosetta Stone, enabling companies to speak the same SIF-prevention language, compare data, and collaborate meaningfully. But EBS integration *should* be tailored. Companies can start in different places, assign roles differently, and adapt visual styles or communication strategies to fit their unique context. EBS brings structure to what already works. Don't replace your safety program; use EBS to give it a SIF-prevention backbone. And when there are opportunities to evolve the science itself, work collaboratively through communities of practice instead of going it alone. As the Kenyan Olympic team put it, a community is "a group of people who agree to grow together." That's how EBS moves forward.

Lesson 4: Invest in Quality Training

Integrating EBS effectively requires a major investment in building knowledge and skills across the organization, from executives to front-line workers. EBS concepts take time and practice to apply, and training must be designed to develop real competency, not just awareness. Because safety professionals are often tapped as internal subject matter experts, their training should come first. Training should also be tailored to each audience, recognizing that different stakeholders play different roles in EBS integration. Field workers need practical application, supervisors need coaching on how to lead EBS efforts, and executives need a clear understanding of their role in sustaining the approach. The most effective programs combine hands-on coaching, real-world application, and ongoing reinforcement to ensure learning translates into practice. Like applying "multiple coats of paint," true competency requires layering knowledge over time. While training must be reinforced and repeated, the real focus should be on quality, ensuring that every level of the organization has the skills and confidence to apply EBS in their daily work.

Lesson 5: Go Slow

EBS is a paradigm shift, not a quick fix. It is a journey that takes years, not months. Rolling it out too fast overwhelms workers and leaders, leading to resistance and frustration. The most effective methods take a deliberate, phased approach, allowing teams to develop competency before layering in more advanced concepts. Some of the best rollouts I've seen took a full year just to introduce the energy wheel before moving to high-energy hazard identification and Direct Controls. While field integration takes time, EBS can be integrated into other areas like design, Safety Walks, metrics, and learning in parallel, ensuring progress without overwhelming front-line workers. Organizations that pace their rollout thoughtfully build stronger, more sustainable programs that workers embrace rather than endure.

Lesson 6: Don't Miss the Middle

Supervisors are the most influential safety leaders in the organization, yet they are often overlooked. Leadership buys in, front-line workers get trained, but without engaged supervisors, EBS collapses. Crews take their cues from supervisors more than anyone else. If supervisors are not fully on board, they will treat EBS as optional, and their teams will follow suit. The best safety programs do not just train supervisors; they empower them. That means giving them more than surface-level exposure. Supervisors should be actively engaged in planning discussions, have a voice in how EBS principles are applied in the work they manage, and be equipped with the tools to lead by example. When field leaders truly understand EBS and believe in its value, their influence cascades through the workforce, turning compliance into commitment. Organizations that prioritize supervisor buy-in create lasting change. Those who neglect it create a program that only works when someone is watching. As my colleague Matt Compher said, whatever training and support you plan to give to field supervisors, triple it.

Lesson 7: Learn from the River, Not the Raindrop

EBS shifts the focus from reacting to isolated incidents to learning from systemic patterns. Shared definitions and classification models like SCL and HECA allow companies to categorize incidents and observations the same way. This consistency makes it possible to combine data across projects and organizations, generating datasets large enough to uncover meaningful and statistically valid trends. It's a decisive break from the old habit of drawing conclusions from one-off events within a single company. Trying to improve safety by reacting to individual incidents is like trying to understand a river by studying a single raindrop. A raindrop tells us nothing about the direction, depth, or strength of the current. But the river does. By analyzing trends in high-energy hazards, Direct Controls, and precursors, organizations can detect systemic weaknesses and direct their efforts where they matter most. Those who learn from the river will build safer, stronger systems than those chasing raindrops.

Lesson 8: Start Simple with the Energy Wheel

With leadership buy-in and resources in place, the next question is where to begin. EBS includes a range of concepts, tools, and methods, making it impractical to integrate everything at once. The most effective starting point is the energy wheel. It is simple, intuitive, and highly effective, improving hazard recognition without adding administrative burden. Because it integrates seamlessly into existing safety practices such as pre-job safety meetings, it delivers immediate value while building a foundation for deeper EBS integration. More importantly, the energy wheel reinforces the fundamental principle of EBS: that energy causes harm. By establishing a common language and structured approach to hazard recognition, it prepares the workforce to systematically assess energy sources, laying the groundwork for high-energy hazard identification and control assessment. When integrated successfully, it not only strengthens hazard recognition but also creates an appetite for deeper engagement with high-energy hazards, controls, and other advanced EBS topics. Organizations that begin with the energy wheel set themselves up for long-term success by ensuring that hazard recognition is comprehensive, consistent, and aligned with an evidence-based approach to safety.

Lesson 9: Controls Are a Tough Concept

Controls are the mechanism through which SIFs are prevented, making them the most critical concept in EBS. Every part of EBS integration, from design to measurement, ultimately ties back to controls, yet assessing and verifying them remains one of the biggest challenges. Even experienced safety professionals often struggle to distinguish between robust, reliable controls and weaker administrative measures. It is even harder to make controls function consistently in the field. Dynamic work environments, operational constraints, and human variability all test their reliability. Companies that succeed in SIF prevention go beyond defining controls, they ensure that controls are understandable, practical, and effective under real-world conditions.

Lesson 10: It Will Take a Long Time to Know When a SIF Prevention Strategy Works

SIFs are exceedingly rare, occurring too infrequently within a single organization to allow for direct statistical validation. Even if SIF rates decline, it will be impossible to isolate and attribute the impact to any one strategy. These rates only stabilize at the industry level, where billions of worker hours are amassed. This uncertainty doesn't mean EBS should be abandoned. It means we must be disciplined in measuring what we can. The best indicator of success is not a fluctuating SIF rate but the presence and effectiveness of controls. Organizations that focus on verifying and improving controls, rather than chasing lagging indicators, will be the ones making meaningful progress in SIF prevention. If enough organizations commit to SIF prevention strategies, their collective impact may eventually shift industry-wide outcomes.

PARTING THOUGHTS

If you've made it to the end of this book, thank you. I hope the ideas have challenged your thinking, clarified a practical path forward, and, most importantly, equipped you to take meaningful action. There's a lot to absorb, especially if it challenges long-held assumptions. But if you leave with a clearer understanding of the state of knowledge and a sharper lens to focus your safety efforts on preventing SIFs, then the book has done its job.

The road ahead won't be easy. The forces we're trying to control are strong, and so are the habits, assumptions, and beliefs we're asking people to reconsider. Changing how we approach safety, especially in organizations that have built entire systems around traditional models, will take time and persistence. We will face resistance, and that's to be expected. But the community working on this problem is strong and dedicated. If we stay grounded in evidence, committed to learning, and open to collaboration, we will succeed.

EBS is not a silver bullet. It's a scientific framework that helps us see, understand, and respond to the conditions that lead to the most devastating outcomes. Because it's rooted in science, it will grow and evolve when challenged by new perspectives and stronger evidence. And that is exactly how it should be. Safety science is not a destination but a discipline, one that grows through collective effort and honest learning.

This book ends here, but the work does not.

Glossary of EBS and SIF Prevention Terms

In the field of safety, definitions play a crucial role in fostering effective collaboration and advancing scientific knowledge. The following definitions underpin the research in EBS and SIF prevention. Many of these definitions are published in peer-reviewed scientific literature, included in standards, and governed by communities of practice.

SERIOUS INJURIES AND FATALITIES (SIFs)

1. **Serious Injury and Fatality (SIF):** An incident that results in a life-ending, life-threatening, or life-altering outcome.
   - 1.1 ***Life Ending*:** Loss of life.
   - 1.2 ***Life Threatening*:** An injury that requires immediate medical intervention or life-saving support to save an injured person's life.
   - 1.3 ***Life Altering*:** An injury where the injured worker will not fully recover and will most likely suffer permanent impairment from the loss of use of a major internal organ, body function, or body part.
2. **Low-Severity Injury:** An injury that does not result in a life-ending, life-threatening, or life-altering outcome.

SCIENCE OF ENERGY

3. **Energy-Based Safety (EBS):** A targeted approach to SIF prevention based in the knowledge that energy causes harm, more energy causes more harm, controls save lives, and people drive performance.
4. **Energy:** The capacity to do work, measured in Joules.
5. **Hazard:** A source of hazardous energy with the potential to cause harm.
6. **Energy Wheel:** A visual tool with ten hazardous energy types organized into a circle to assist with the comprehensive identification of hazards.
7. **High-Energy Hazard:** A hazard involving at least 1500 Joules of available energy, where contact with the human body is most likely to result in a Serious Injury or Fatality (SIF).
8. **Low-Energy Hazard:** A hazard involving less than 1500 Joules of available energy, where contact with the human body is most likely to result in a Low-Severity Injury (LSI).
9. **High-Energy Icons:** A field-ready simplification of high-energy computations for hazard types that are almost always above the 1500 Joule threshold.
10. **STKY:** An acronym for the Stuff That Kills You, a field-friendly synonym for high-energy hazards.

CONTROLS

11. **Energy Control:** Timely, tangible, and targeted measures that actively mitigate or neutralize hazardous energy.
    11.1 ***Timely*:** Actively in place and functioning when a high-energy hazard is present.
    11.2 ***Tangible*:** Physically present and observable on the worksite.
    11.3 ***Targeted*:** Specifically designed and installed to address a specific high-energy hazard.
12. **Direct Control:** An energy control that mitigates high energy and works even if someone makes a mistake.
13. **Alternative Control System:** An error-mitigating set of controls that have at least two complementary energy controls from at least two of the following categories: physical obstacle, dedicated monitoring, and visual reminder.
    13.1 ***Physical Obstacle*:** An obstruction that blocks the path or hinders progress toward a high-energy hazard.
    13.2 ***Dedicated Monitoring*:** Devoted and continuous attention to a high-energy hazard through a human or technological system.
    13.3 ***Visual Reminder*:** A visible warning of the presence of a high-energy hazard.

HUMAN FACTORS

14. **Error:** A deviation from intended actions or decisions, which can result in unintended outcomes.
15. **Precursor:** Identifiable event, condition, or action that serves as a warning sign of uncontrolled high-energy exposure or incident.

SAFETY CLASSIFICATION AND LEARNING (SCL) MODEL

16. **High-Energy Incident:** Instance where the high-energy source was released and where the worker came in contact with or proximity to the high-energy source.
    16.1 ***Energy Release*:** Instance in time when there is a temporary loss of control of the high energy, or when the high-energy changes state in the work environment.
    16.2 ***Contact*:** Instance when the high energy is transmitted to the human body.
    16.3 ***Proximity*:** Condition where the boundary of the high-energy exposure is within 6 feet of the worker, within a confined space, or within a situation where a worker cannot escape the energy source (restricted egress).
17. **High-Energy Serious Injury or Fatality (HSIF):** Incident with a release of high energy where a serious injury is sustained.
18. **Low-Energy Serious Injury or Fatality (LSIF):** Incident with a release of low energy where a serious injury is sustained.

19. **Potential Serious Injury or Fatality (PSIF):** Incident with a release of high energy in the absence of a Direct Control where a serious injury is not sustained.
20. **Capacity:** Incident with a release of high energy in the presence of a Direct Control where a serious injury is not sustained.
21. **Exposure:** Condition where a high-energy hazard does not have a corresponding Direct Control.
22. **Success:** Condition where a high-energy hazard does have a corresponding direct control.
23. **Low-Severity:** Low-energy incidents or conditions that did not result in, and are unlikely to result in, a SIF.

KEY SAFETY ACTIVITIES

24. **Prevention through Design (PtD):** Process of changing the features of the work product so it is safer to build, operate, and maintain.
25. **Pre-Job Safety Brief:** A safety-focused discussion that takes place on site immediately before work begins. Also referred to as a pre-task plan, toolbox talk, tailgate talk, or tailboard.
26. **Safety Walk:** On-site activity where leaders observe active work and engage with employees to understand, assess, and improve safety conditions.
27. **Energy-Based Observation (EBO):** A safety walk that is focused on the identification and assessment of high-energy hazards and their corresponding controls.
28. **High-Energy Control Assessment (HECA):** The proportion of high-energy hazards with a corresponding Direct Control (expressed as a percentage).

Appendix 1

Example Serious Injury and Fatality (SIF) Criteria

The following criteria describe life-ending, life-threatening, and life-altering injuries that are considered "serious." The criteria were adapted with permission from the Edison Electric Institute (EEI) SIF Criteria published 1 January 2025.[1] The EEI source document includes footnotes, explanations, and frequently asked questions (FAQs) that provide further clarification.

1. Fatalities
2. Amputations (involving bone) excluding distal phalanx unless thumb, index, or great toe.
3. Head trauma that results in a traumatic brain injury (TBI), intracranial bleeding, or loss of consciousness for greater than 30 minutes.
4. Injury or trauma to vital organs including brain, spinal cord, heart, lungs, kidneys, liver, spleen, large and small intestine, and stomach.
5. Bone fractures requiring surgery for repair (e.g., pins, rods, screws, or plates) but excludes fingers and toes.
6. Acute traumatic herniated disc with neurologic sensory or motor deficit.
7. Second-degree burn to 10% or more of the body surface; third-degree burn to 5% or more of body surface, or third-degree burn requiring skin graft.
8. Eye injuries resulting in permanent vision loss or change in vision.
9. High pressure injection injuries requiring surgical debridement and irrigation.
10. Heat stroke.
11. Dislocation of the hip, elbow, or knee excluding the patella (kneecap).
12. Electrical contact injuries resulting in permanent loss of function of a joint, cardiac dysrhythmia, or requiring surgical repair, skin grafting, or amputation.
13. Vascular trauma requiring surgery.
14. Acute chemical or radiological exposure resulting in injury to vital organs including brain, spinal cord, heart, lungs, kidneys, liver, spleen, large and small intestine, and stomach.
15. Other injuries that meet the life-threatening or life-altering definitions.

NOTE

1 Edison Electric Institute. (2024). Serious Injury and Fatality Criteria: Approved Changes. Retrieved April 18, 2025, from https://www.eei.org/issues-and-policy/power-to-prevent-sif.

Appendix 2

Guide to Physical Energy Computations

For most physical hazards, the magnitude of energy can be estimated from a set of basic parameters such as height, mass, and velocity. This appendix includes the physics equations and examples for gravity, motion, electrical, pressure, and mechanical hazards. Other hazard types, such as chemical and radiation, do not lend themselves to the same physical energy computations and, instead, require different assessments where dosage is measured against permissible exposure limits (PELs) published in associated standards.

Each energy source is described, equations are noted, and two examples are provided, one for high energy (1500 Joules or above) and one for low energy (below 1500 Joules). The computations are performed in metric units with rough equivalence in Imperial units in parentheses.

Please note: The energy computations provided in this book are intended solely to illustrate how the research was conducted and to offer examples of how energy magnitude can be estimated in practical contexts. These examples are not comprehensive, nor are they intended to represent all calculation methods applicable to the assessment of hazardous energy. They should not be used as a substitute for detailed engineering analysis where precision, accuracy, and completeness are required.

GRAVITY

Gravitational energy is the potential energy stored in an object due to its height above a reference point. It is determined by three factors: the object's mass, the vertical distance from the reference point (usually the ground), and the gravitational constant. In safety contexts, gravitational energy becomes a concern when it is released like when an object or person falls. As the object descends, gravitational potential energy is converted into kinetic energy (motion). Injuries can result either from a person falling or from a falling object striking someone below.

Gravitational energy (E) is directly proportional to the mass of an object or individual, their height above a reference point, and the gravitational constant denoted as G. In the International System of Units (SI), mass is quantified in kilograms (kg), height in meters (m), and the gravitational constant (G) is standardized at 9.8 m/s^2.

$$E = \text{mass} \times \text{height} \times \text{gravitational constant}$$

Examples

- 0.45 kg (1 lb) tape measure falls on a worker from 3 m (10 ft)

  $$E = 0.45\,kg \times 3\,m \times 9.8\,m/s^2 = 15\,Joules$$

  $$E = 15\,Joules \times 0.74 \approx 10\,foot\text{-}pounds$$

  Conclusion: Low energy

- 90 kg (200 lb) falls 4.6 m (15 feet)

  $$E = 90\,kg \times 4.6\,m \times 9.8\,m/s^2 = 4{,}057\,Joules$$

  $$E = 4{,}057\,Joules \times 0.74 \approx 3{,}000\,foot\text{-}pounds$$

  Conclusion: High energy

MOTION

Motion energy, formally called kinetic energy, is the energy an object has because it's moving. This type of energy applies to objects moving in a straight line, like a rolling cart or a sliding tool. It doesn't include energy from spinning, stretching, or squishing, which are classified here as mechanical energy.

How much kinetic energy an object has depends on:

- How heavy it is (its mass), and
- How fast it's moving (its velocity).

The faster something moves, the more energy it has, but the relationship isn't linear, it's squared. For example, you double the speed, the energy increases by four times. This means even small increases in speed can make a big difference in the severity of an incident.

In the International System of Units (SI), mass is quantified in kilograms (kg), and velocity is measured in meters per second (m/s). In Imperial units, mass is represented in pounds (lbs), and velocity is expressed in miles per hour (mph). As a point of reference, 1 m/s is equivalent to 3.6 kilometers per hour (kph) or 2.2 miles per hour (mph).

$$E = 0.5 \times mass \times velocity^2$$

Examples

- Workers carrying a 100 kg (220 lb) pipe strike the torso of another worker at 5 kph (3 mph). 5 kph is equal to 1.34 m/s.

  $$E = 0.5 \times 100\,kg \times (1.34\,m/s)^2 = 90\,Joules$$

  $$E = 90\,Joules \times 0.74 \approx 66\,foot\text{-}pounds$$

  Conclusion: Low energy

- 1200 kg (2646 lbs) vehicle strikes worker at 40 kph (25 mph). 40 kph is equal to 11 meters per second.

$$E = 0.5 \times 1200\,\text{kg} \times \left(11\,\text{m/s}\right)^2 = 72{,}600\,\text{Joules}$$

$$E = 72{,}600\,\text{Joules} \times 0.74 \approx 53{,}000\,\text{foot-pounds}$$

Conclusion: High energy

ELECTRICAL

Electrical energy becomes a serious hazard when it enters the human body as current. Unlike gravitational or motion energy, which act externally, electrical energy passes through the body itself. As it travels, it is converted into heat, disrupting the body's internal systems and causing injury to tissues, nerves, and organs.

In practical terms, most electrical injuries occur when a person contacts an energized object or circuit. The severity of the injury depends on how much current flows, how long the contact lasts, and which path the current takes through the body. For estimation purposes, we assume a body resistance of 1500 ohms and treat all the electrical energy as being converted into heat. This simplification helps standardize the calculation and makes the energy-based framework easier to apply consistently.

One of the most important characteristics of electrical energy is that it increases rapidly with voltage. Doubling the voltage doesn't just double the energy, it increases it by a factor of four. This exponential relationship makes even brief contacts with high-voltage sources potentially life-threatening.

Electrical energy relies on time measured in seconds (s), voltage (V), amperage (A), the assumed resistance of the human body (1500 ohms), and the assumption that all electrical energy transforms into heat. These computations remain consistent regardless of whether SI or imperial units are employed.

$$E = \text{time} \times \text{voltage}^2 / \text{resistance}$$

Examples

- Worker touches a 220 V wire for 2 seconds

$$E = 2\,\text{s} \times 220\,\text{V}^2 / 1500\,\text{ohm} = 64.6\,\text{Joules}$$

$$E = 64.6\,\text{Joules} \times 0.74 \approx 48\,\text{foot-pounds}$$

Conclusion: Low energy

- Electrical contact for 0.05 seconds with a 10 kV circuit breaker

$$E = 0.05\,s^{*}\,10{,}000\,V^2/1{,}500\,ohm = 3{,}333\ Joules$$

$$E = 3{,}333\ Joules \times 0.74 \approx 2{,}500\ foot\text{-}pounds$$

Conclusion: High energy

PRESSURE

Pressure energy is stored in compressed gases or liquids often contained in vessels and pipes such as tanks, compressed gas cylinders, pipelines, and hoses. When pressure is suddenly released, it can cause violent bursts, flying projectiles, or rapid jetting, all of which pose serious risks to workers nearby.

The amount of pressure energy depends on two main factors:

- The pressure inside the container (how tightly the gas or liquid is compressed), and
- The volume of the container (how much compressed material is stored).

Both variables matter equally. A small container at very high pressure can hold as much energy as a larger one at lower pressure. That's why pressure hazards can be difficult to judge by size alone, they require calculation.

Internal pressure typically measured in pounds per square inch (psi), with 1 psi equivalent to 7000 Pascals (Pa). When using the metric system, we measure the volume of a vessel in liters where 1 liter corresponds to approximately 0.264 gallons. In the case of linear vessels such as pipes and hoses, we measure the energy per meter by estimating the pressure within the pipeline (psi) and the diameter of the pipe (m). For reference, 1 Joule equals to 1 Pascal*m^3.

$$\text{For Vessels}: E\left(Pa^{*}\,m^3\right) = 7000 \times \text{pressure}\left(\text{in psi}\right) \times 0.001 \times \text{volume}\left(\text{in L}\right)$$

$$\text{For Pipes}: E = 7000 \times \text{pressure}\left(\text{in psi}\right) \times \pi\left(0.5 \times \text{diameter}\left(\text{in m}\right)\right)^2$$

EXAMPLES

- Working near a 2-inch (5 cm) natural gas line at 40 psi (275 kPa)

$$E = 7000 \times 40\,psi \times 3.14 \times \left(0.5 \times 0.05\right)^2 = 550\ Joules$$

$$E = 550\ Joules \times 0.74 \approx 400\ foot\text{-}pounds$$

Interpretation: Low energy

- Welding with a 2.5 Gallon (10 liter) acetylene cylinder at 250 psi (1724 kPa)

$$E = 7{,}000 \times 250\,\text{psi} \times 0.001 \times 10\,\text{L} = 17{,}500\,\text{Joules}^1$$

$$E = 17{,}500\,\text{Joules} \times 0.74 \approx 12{,}950\,\text{foot-pounds}$$

Interpretation: High energy

Note: In these equations and in the associated energy severity assessment tool, pressure is estimated for SI units in pounds per square inch (psi) rather than kilopascals (KPa). This convention has been used because psi is the typical convention used in most industrial applications. Pounds per square inch may be converted to kPa at 1 psi = 6.89 kPa.

MECHANICAL

Mechanical energy typically takes one of two primary forms: rotational energy (E_r) and elastic energy (E_e). Rotational energy involves spinning components such as grinders, turbines, gears, or pulleys, while elastic energy is in objects under tension or compression such as a cable in a pulley.

Computing rotational energy (E_r) involves two key parameters: the object's rotational inertia (I), influenced by both its mass and shape, and the angular velocity. Rotational inertia (I) is measured as $I = 0.33 \times \text{weight} \times \text{length}^2$ for a rod or $I = 0.5 \times \text{weight} \times \text{radius}^2$ for a cylinder. For a rod, this assumes rotation about one end. For a cylinder, this assumes rotation along the z axis. Angular velocity is measured in radians per second where 1 rotation per minute (rpm) equates to approximately 0.104 radians per second.

Elastic energy (E_e), on the other hand, depends on the stiffness of the object (k) measured in Newtons per meter. For example, a spring that extends by 10 cm when supporting 85 kg (830 Newtons) has a stiffness of k = 8300 N/m. The distance in meters is the difference between the rest length and the current length.

$$\text{For rotation}: E_r = 0.5 \times I \times \text{angular velocity}^2$$

$$\text{For tension or compression}: E_e = 0.5 \times k \times \text{distance}^2$$

EXAMPLES

- A grinder wheel with a 0.114 m (4.5-inch) diameter weighing 300 grams is rotating at a speed of 11,000 RPM (1,151 rad/s).

$$I\left(\text{kg} \times \text{m}^2\right) = 0.5 \times 0.3\,\text{kg} \times \left(0.114\text{m}/2\right)^2 = 0.00049\,\text{kg} \times \text{m}^2$$

$$E_r = 0.5 \times 0.00049\,\text{kg} \times \text{m}^2 \times (1{,}151\,\text{rad/s})^2 = 320\,\text{Joules}$$

$$E_r = 320^{*}\,0.74 \approx 237\,\text{foot-pounds}$$

Conclusion: Low energy

- Synthetic sling stretches 0.40 m (16 inches) while supporting 1,134 kg (2,500 lbs). 1,134 kg is equal to 11,113 Newtons.

$$k = 11{,}113\,\text{N} / 0.40\,\text{m} = 27{,}800\,\text{N/m}$$

$$E_e = 0.5^{*}\,27{,}800\,\text{N/m}^{*}\,(0.40\,\text{m})^2 = 2{,}224\,\text{Joules}$$

$$E_e = 2{,}224\,\text{Joules} \times 0.74 \approx 1{,}640\,\text{foot-pounds}$$

Conclusion: High energy

Appendix 3

Example Tasks with High-Energy Hazards and Direct Controls

This appendix presents common high-energy hazards with example tasks where they may occur, along with the Direct Controls typically used to manage them. These examples support understanding of energy-based safety (EBS) and help practitioners recognize how serious hazards often appear in routine work. While not exhaustive, the lists illustrate how EBS principles can be applied in practice.

Each section includes example tasks and controls associated with a specific high-energy hazard. While common Direct Controls are listed, every control must be evaluated in context to ensure it meets the definition: a targeted safeguard that mitigates high energy and remains effective even in the presence of human error. All controls must also be properly installed, verified, and used. Additional guidance on evaluating controls is provided in Chapter 5.

Whenever possible, high-energy hazards should be eliminated or designed out by avoiding proximity to them. This represents the highest-order control. Although not always feasible, elimination should remain a primary consideration during planning and design. The examples in this appendix assume the hazard could not be eliminated and must therefore be controlled during execution.

Some Direct Controls, such as physical barriers, apply broadly across many high-energy hazards. Others, like radiation shielding, are specific to particular hazard types. The following two controls apply across nearly all hazard types and are marked with an asterisk (*) throughout the appendix:

- *Hard physical barriers*, which prevent workers from entering areas where high-energy hazards are present.
- *Remote operations or automation*, which remove the need for workers to be physically present near the hazard.

1. Suspended Load

High-Energy Hazard Examples

- Placing steel beams with rigging and crane
- Installing valves and fittings with chain hoist and beam trolley
- Loading and unloading heavy materials from a flatbed truck
- Lifting insulator bundles with synthetic slings and bucket truck
- Installing switchgear cabinets with pallet forks and telehandler
- Setting structural tower sections with slings and a crane
- Climbing a tower with a heavy tool
- Hoisting air-handling units onto rooftops using a crane and rigging gear
- Positioning cable reels with chainfall and A-frame
- Installing turbine blades with fixture and overhead gantry crane
- Positioning pipeline sections with spreader bar and crane
- Lifting concrete buckets with a tower crane

Direct Control Examples

- Ratchet straps or tie-downs for material security
- Tool tether for tools at height

* Hard physical barriers that prevent worker access to unsafe areas
* Remote operations or automation

2. Fall from Elevation (Greater than 4 Feet)

High-Energy Hazard Examples

- Connecting steel beams while standing on a structural frame
- Setting precast concrete panels while working on scaffolding
- Installing overhead power lines while working in an elevated bucket
- Securing roofing materials while working on a sloped roof
- Welding pipe supports while working on a pipe rack
- Inspecting transmission towers while climbing a lattice structure
- Pouring concrete into formwork while standing on a bridge deck edge
- Adjusting lighting fixtures while working on a mobile platform
- Repairing air-handling units while working on a rooftop
- Bolting turbine components while standing on the upper platform of a wind turbine
- Cutting tree limbs while working from a bucket truck
- Aligning large ductwork from an aerial lift

Direct Control Examples

- Fall arrest system
- Fall restraint system
- Covers over holes

* Hard physical barriers that prevent worker access to unsafe areas

* Remote operations or automation

3. Mobile Equipment or Traffic with Workers on Foot

High-Energy Hazard Examples

- Excavating trenches with a hydraulic excavator while workers set pipe nearby
- Directing traffic with a flagger while vehicles pass through a work zone
- Grading a roadway with a motor grader while surveyors take measurements
- Moving materials with a forklift while workers position loads
- Installing guardrails while working on the shoulder of an active highway
- Operating a dump truck while laborers work near the haul route
- Lifting steel beams with a crane while riggers guide the load
- Operating a concrete mixer truck while crews handle chutes and pour concrete
- Moving wind turbine components with a multi-axle transporter while workers assist
- Backfilling around foundations with a loader while workers compact soil
- Operating a rail-mounted crane while ground crews position track sections
- Repairing pavement markings while working from a slow-moving truck-mounted platform

Direct Control Examples

- Automatic shutoff system that eliminates the potential for collision between equipment and workers
- Hard physical barriers that stop progression of equipment into unsafe areas (e.g., concrete barrier)

* Hard physical barriers that prevent worker access to unsafe areas
* Remote operations or automation

4. Motor Vehicle Operation Over 30 mph (Vehicle Occupant)

High-Energy Hazard Examples

- Driving a pickup truck between job sites on a highway
- Transporting workers in a van on a rural road
- Hauling equipment in a flatbed truck on a construction site access road
- Operating a dump truck at a mine or quarry
- Driving a utility bucket truck between power line maintenance locations
- Transporting materials in a utility task vehicle (UTV) on a pipeline right-of-way
- Operating an all-terrain vehicle (ATV) on rough terrain to inspect remote work areas
- Driving a snowmobile to access a winter job site
- Operating a water truck on an unpaved road for dust control
- Driving a semi-truck carrying oversized loads for a wind farm project
- Operating a personnel transport bus on a large industrial site
- Conducting a high-speed response in a fire truck to an emergency at a worksite

Direct Control Examples

- Highway-rated vehicle (includes steel frame, seat belts, airbags, etc.)
- Non-highway-rated vehicles with rollover protection and seat belts

* Remote operations or automation

5. Heavy Rotating Equipment

High-Energy Hazard Examples

- Working near a large turbine in a power system
- Maintaining a heavy flywheel in industrial equipment
- Working near a high-speed pump in a processing facility
- Operating a rock crusher with massive spinning drums
- Working around a rotating drum in a manufacturing process
- Maintaining a large compressor with spinning blades
- Operating a wind turbine with a rotating hub
- Working on large rollers in a material processing system
- Servicing a high-speed fan in a cooling system
- Operating a large boring machine for heavy-duty drilling
- Repairing a rotating drive shaft on industrial equipment
- Aligning a spinning drum in a drying or mixing system

Direct Control Examples

- De-energization with lock-out tag-out (LOTO) applied
- Machine guards
- Interlocks

* Hard physical barriers that prevent worker access to unsafe areas
* Remote operations or automation

6. High Surface Temperatures (Greater than 150° Fahrenheit)

High-Energy Hazard Examples

- Welding steel components during structural fabrication
- Grinding or cutting metal with an angle grinder after sustained use
- Applying bitumen or asphalt during paving operations
- Working near engine manifolds during heavy equipment maintenance
- Removing cast parts from an industrial foundry mold
- Operating a heat-treated pipe bending system in a fabrication yard
- Handling materials inside concrete curing enclosures
- Replacing refractory insulation inside high-temperature furnaces
- Clearing blockages from heated extrusion equipment in manufacturing
- Performing thermal spraying (metal coating) on structural components
- Maintaining industrial ovens or kilns during refractory replacement
- Working near heated molds in plastic or rubber manufacturing

Direct Control Examples

- Automatic fuel shutoff system
- Interlocks
- Rated heat resistant clothing with no skin exposed
- Thermal surface insulation

* Hard physical barriers that prevent worker access to unsafe areas

* Remote operations or automation

7. Steam

High-Energy Hazard Examples

- Opening valves on live steam lines during plant operations
- Installing or removing insulation on active or recently isolated steam piping
- Repairing steam traps or condensate return systems
- Replacing gaskets on steam flanges in a power plant
- Conducting leak checks on high-pressure steam systems
- Blowing down steam boilers during normal maintenance
- Draining steam lines for shutdown or reconfiguration
- Maintaining autoclaves or steam sterilization equipment
- Working near steam headers in a turbine hall during startup
- Clearing blockages in steam-heated process lines
- Disassembling steam-driven pumps for internal inspection
- Inspecting or calibrating instruments near live steam systems

Direct Control Examples

- De-energization with lock-out tag-out (LOTO) applied
- Energy isolation (e.g., squeezing and blocking; closing valves)
- Automatic shutoff system
- Rated steam suits with no skin exposed

* Hard physical barriers that prevent worker access to unsafe areas
* Remote operations or automation

8. Fire with a Sustained Fuel Source

High-Energy Hazard Examples

- Welding slag contacting flammable clothing or materials
- Working in proximity to an active wildfire or prescribed burn
- Disturbing natural gas lines near powered equipment or metal hand tools
- Using cutting torches near combustible construction materials
- Performing hot work in a confined space with residual flammable vapors
- Operating spark-producing tools near open containers of solvents or fuels
- Refueling equipment near ignition sources without proper grounding or bonding
- Hot tapping into active hydrocarbon pipelines
- Improper storage or handling of flammable aerosols in maintenance shops
- Grinding or metal cutting in areas with accumulated sawdust or wood shavings
- Repairing or modifying tanks that previously held combustible liquids
- Performing electrical work near combustible vapor zones without classified equipment

Direct Control Examples

- De-energization with lock-out tag-out (LOTO) applied
- Energy isolation (e.g., squeezing and blocking; closing valves)
- Fuel source separation
- Automatic fire suppression system
- Automatic fuel shutoff system

* Hard physical barriers that prevent worker access to unsafe areas
* Remote operations or automation

9. Explosion

High-Energy Hazard Examples

- Using a cutting torch on a sealed drum that previously held flammable liquids
- Hot work in an area with accumulated flammable gas or vapor
- Opening process equipment that may contain pressurized flammable gases
- Cleaning grain silos or mills where combustible dust is suspended in air
- Using non-intrinsically safe tools in classified hazardous locations
- Charging or discharging batteries in poorly ventilated enclosures
- Adding incompatible chemicals in a mixing or treatment process
- Purging or commissioning gas lines without proper venting
- Conducting blasting operations in mining or quarry environments
- Servicing acetylene cylinders or regulators without proper procedures
- Using portable fuel containers near ignition sources in enclosed spaces
- Repairing tanks that previously held volatile liquids without full cleaning and testing

Direct Control Examples

- De-energization with lock-out tag-out (LOTO) applied
- Automatic fire suppression system
- Blast blankets
- Energy isolation (e.g., squeezing and blocking; closing valves)
- Explosion suppression system
- Fuel source separation
- Mechanical pressure release devices
- Rated panel enclosures
- Hard physical barriers that prevent worker access to unsafe areas
- Remote operations or automation

10. Excavation or Trench with a Depth of 5 ft or Greater

High-Energy Hazard Examples

- Installing utility piping in a trench deeper than five feet without protective systems
- Excavating footings for structural foundations near adjacent loads or structures
- Laying stormwater culverts in an unshored trench
- Repairing water mains in narrow, vertically cut excavations
- Installing electrical conduit in deep trenches using hand digging tools
- Setting manholes or vaults inside excavations exceeding five feet in depth
- Tying rebar for below-grade walls in unsupported excavations
- Jetting or vacuuming around utilities in deep trenches without sloping or shielding
- Excavating in wet or loose soil without proper trench support
- Placing formwork or concrete in a trench with standing water and unstable walls
- Performing work below trench boxes or outside the protected zone
- Hand tunneling or pipe jacking from within a deep access pit

Direct Control Examples

- Excavation support system (e.g., sheet piles)
- Trench box
- Sloping or benching

* Hard physical barriers that prevent worker access to unsafe areas
* Remote operations or automation

11. Electrical Contact with 50 Volts or Greater

High-Energy Hazard Examples

- Installing light fixtures on an energized circuit without lockout
- Testing live circuits in an electrical panel with the cover removed
- Running conduit near exposed wiring during renovation work
- Splicing energized conductors in a junction box
- Working in a lift near overhead power lines without proper clearance
- Replacing outlets or switches in energized systems without verifying de-energization
- Performing maintenance on HVAC systems without disconnecting power
- Drilling or cutting into walls with hidden energized wiring
- Connecting or disconnecting temporary power at a construction site
- Installing or troubleshooting control panels in live equipment
- Using extension cords with damaged insulation in wet conditions
- Working in an electrical vault or switchgear room without verified isolation
- Performing maintenance inside a machine room without isolating power

Direct Control Examples

- De-energization with lock-out tag-out (LOTO) applied
- Equipotential zones (EPZ)
- Electrical insulation barriers (e.g., rubber cover)
- Specialty Personal Protective Equipment (PPE) such as insulating gloves, sleeves, and dielectric footwear that provide full coverage

* Hard physical barriers that prevent worker access to unsafe areas

* Remote operations or automation

12. Arc Flash

High-Energy Hazard Examples

- Removing covers from energized switchgear or motor control centers
- Testing circuits inside 480-volt electrical panels
- Installing or removing fuses in high-capacity disconnects
- Racking in or out circuit breakers in energized switchgear
- Working on bus bars in industrial control panels
- Resetting breakers after a known electrical fault
- Performing infrared scans on energized electrical equipment
- Operating disconnects or breakers in high-energy systems
- Installing current transformers on energized conductors
- Conducting live maintenance on transformers or switchboards
- Modifying wiring inside panels energized above 240 volts
- Accessing utility-owned service gear during commissioning or startup

Direct Control Examples

- De-energization with lock-out tag-out (LOTO) applied
- Blast blankets
- Equipotential zones (EPZ)
- Rated panel enclosure
- Arc-rated PPE with no skin exposed

* Hard physical barriers that prevent worker access to unsafe areas

* Remote operations or automation

13. High Dose of Toxic Chemical or Radiation

High-Energy Hazard Examples

- Spraying industrial paint in enclosed or poorly ventilated areas
- Using strong cleaning chemicals during equipment maintenance
- Pouring or transferring large amounts of ammonia or bleach-based products
- Removing old paint or coatings that contain lead
- Cutting or welding metal parts with hazardous surface coatings
- Cleaning tanks that previously held harmful chemicals
- Removing old insulation materials during building renovation
- Using equipment that contains radioactive material to inspect components
- Taking X-ray images of welds or parts using portable radiation sources
- Working near radioactive materials in research or medical settings
- Applying pesticides or soil treatments over a large area
- Cleaning up spilled chemicals from containers or trucks

Direct Control Examples

- De-energization with lock-out tag-out (LOTO) applied
- Automatic shutoff system
- Energy isolation (e.g., closing valves)
- Radiation shielding
- Specialized rated containers
- Local exhaust ventilation
- Chemical or radiation rated PPE

* Hard physical barriers that prevent worker access to unsafe areas
* Remote operations or automation

Appendix 4

Example Safety Classification and Learning (SCL) Model Cases

The following three cases assist with interpreting and applying the SCL model, three detailed example cases are provided. More cases are available in the Edison Electric Institute (EEI) report.[1] The following examples are provided in imperial units to match the EEI report.

CASE 1

An employee was on the top of a de-energized transformer at 25 feet of height with a proper fall arrest system. While working she tripped on a lifting lug and fell from the unguarded edge. She was caught by her fall arrest system, which functioned as designed. The employee sprained her wrist in the fall, which was treated on site by a medical professional.

Classification Decisions

1. ***Was high-energy present?*** Yes, the worker was at 25 feet of height, which exceeds 4 ft of height (see icon)
2. ***Was there a high-energy incident?*** Yes, the worker tripped and fell
3. ***Was a serious injury sustained?*** No, a sprained wrist is not considered serious per the EEI Serious Injury and Fatality (SIF) Criteria
4. ***Was a Direct Control present?*** Yes, a proper fall arrest system was used, which reduces energy exposure to below the high-energy threshold.

Conclusion: Capacity

CASE 2

An employee was working alone and placed an extension ladder against the wall. When he reached 10 feet of height, the ladder feet slid out and he fell with the ladder to the floor. The employee was taken to the hospital for a bruise to his right leg and remained off duty for three days.

Classification Decisions

1. ***Was high-energy present?*** Yes, the worker was at 10 feet of height, which exceeds the 4 ft threshold (see icon)
2. ***Was there a high-energy incident?*** Yes, the energy was released when the worker fell.
3. ***Was a serious injury sustained?*** No, the injury does not meet the EEI SIF criteria.
4. ***Was a Direct Control present?*** No, there were no controls that meet the Direct Control requirements.

Conclusion: PSIF

CASE 3

A crew was closing a 7-ton door on a coal crusher. As the door was lowered, an observer noticed that the jack was not positioned correctly and could tip. The observer also noted that workers were nearby, within 4 feet of the jack.

Classification Decisions

1. ***Was high-energy present?*** Yes, the 7-ton door far exceeds the threshold for gravity per energy calculations
2. ***Was there a high-energy incident?*** No, the observer intervened before the energy was released.
3. ***Was a serious injury sustained?*** No.
4. ***Was a Direct Control present?*** No, the jack was not installed and used properly.

Conclusion: Exposure

NOTE

1 Hallowell, M.R. (2024). *Safety Classification and Learning Model* [technical report]. Edison Electric Institute.

Index

Pages in *italics* refer to figures and pages in **bold** refer to tables.

E

F

G

H

T

www.ingramcontent.com/pod-product-compliance
Lightning Source LLC
Chambersburg PA
CBHW051214031125
PP17305400002B/14
9781041076339